Particle Physics: A Very Short Introduction

VERY SHORT INTRODUCTIONS are for anyone wanting a stimulating and accessible way into a new subject. They are written by experts, and have been translated into more than 45 different languages.

The series began in 1995, and now covers a wide variety of topics in every discipline. The VSI library currently contains over 750 volumes—a Very Short Introduction to everything from Psychology and Philosophy of Science to American History and Relativity—and continues to grow in every subject area.

Very Short Introductions available now:

ABOLITIONISM Richard S. Newman
THE ABRAHAMIC RELIGIONS
  Charles L. Cohen
ACCOUNTING Christopher Nobes
ADDICTION Keith Humphreys
ADOLESCENCE Peter K. Smith
THEODOR W. ADORNO
  Andrew Bowie
ADVERTISING Winston Fletcher
AERIAL WARFARE Frank Ledwidge
AESTHETICS Bence Nanay
AFRICAN AMERICAN HISTORY
  Jonathan Scott Holloway
AFRICAN AMERICAN RELIGION
  Eddie S. Glaude Jr
AFRICAN HISTORY John Parker and
  Richard Rathbone
AFRICAN POLITICS Ian Taylor
AFRICAN RELIGIONS
  Jacob K. Olupona
AGEING Nancy A. Pachana
AGNOSTICISM Robin Le Poidevin
AGRICULTURE Paul Brassley and
  Richard Soffe
ALEXANDER THE GREAT
  Hugh Bowden
ALGEBRA Peter M. Higgins
AMERICAN BUSINESS HISTORY
  Walter A. Friedman
AMERICAN CULTURAL HISTORY
  Eric Avila
AMERICAN FOREIGN RELATIONS
  Andrew Preston
AMERICAN HISTORY Paul S. Boyer

AMERICAN IMMIGRATION
  David A. Gerber
AMERICAN INTELLECTUAL
  HISTORY
  Jennifer Ratner-Rosenhagen
THE AMERICAN JUDICIAL
  SYSTEM Charles L. Zelden
AMERICAN LEGAL HISTORY
  G. Edward White
AMERICAN MILITARY HISTORY
  Joseph T. Glatthaar
AMERICAN NAVAL HISTORY
  Craig L. Symonds
AMERICAN POETRY David Caplan
AMERICAN POLITICAL HISTORY
  Donald Critchlow
AMERICAN POLITICAL PARTIES
  AND ELECTIONS L. Sandy Maisel
AMERICAN POLITICS
  Richard M. Valelly
THE AMERICAN PRESIDENCY
  Charles O. Jones
THE AMERICAN REVOLUTION
  Robert J. Allison
AMERICAN SLAVERY
  Heather Andrea Williams
THE AMERICAN SOUTH
  Charles Reagan Wilson
THE AMERICAN WEST
  Stephen Aron
AMERICAN WOMEN'S HISTORY
  Susan Ware
AMPHIBIANS T. S. Kemp
ANAESTHESIA Aidan O'Donnell

Available soon:

For more information visit our website

www.oup.com/vsi/

Frank Close

PARTICLE PHYSICS

A Very Short Introduction

SECOND EDITION

OXFORD
UNIVERSITY PRESS

Great Clarendon Street, Oxford, OX2 6DP,
United Kingdom

Oxford University Press is a department of the University of Oxford.
It furthers the University's objective of excellence in research, scholarship,
and education by publishing worldwide. Oxford is a registered trade mark of
Oxford University Press in the UK and in certain other countries

Published in the United States of America by Oxford University Press
198 Madison Avenue, New York, NY 10016, United States of America

British Library Cataloguing in Publication Data

Data available

Library of Congress Control Number: 2023942619

ISBN 978-0-19-287375-0

Printed by Integrated Books International, United States of America

Contents

Foreword

We are made of atoms. With each breath you inhale 10 billion trillion atoms of oxygen, which gives some idea of how small each one is. All of them, together with the carbon atoms in your skin, and indeed everything else on Earth, were cooked in a star some 5 billion years ago. So you are made of stuff that is as old as the planet, one-third as old as the universe, though this is the first time that those atoms have been gathered together such that they think that they are you.

Particle physics is the subject that has shown how matter is built and which is beginning to explain where it all came from. It has revealed the existence of a ubiquitous essence, known as the 'Higgs field', which gives mass to matter's fundamental particles. In huge accelerators, several miles in length, we can speed pieces of atoms, particles such as electrons and protons, or even exotic pieces of antimatter, and smash them into one another. In so doing we are creating for a brief moment in a small region of space an intense concentration of energy, which replicates the nature of the universe as it was within a split second of the original Big Bang. Thus we are learning about our origins.

Discovering the nature of the atom 100 years ago was relatively simple: atoms are ubiquitous in matter all around, and teasing out their secrets could be done with apparatus on a table top.

Investigating how matter emerged from Creation is another challenge entirely. There is no Big Bang apparatus for purchase in the scientific catalogues. The basic pieces that create the beams of particles, speed them to within an iota of the speed of light, smash them together, and then record the results for analysis all have to be made by teams of specialists. That we can do so is the culmination of a century of discovery and technological progress. It is a big and expensive endeavour, but it is the only way that we know to answer such profound questions. In the course of doing so, unexpected tools and inventions have been made. Antimatter and sophisticated particle detectors are now used in medical imaging; data acquisition systems designed at CERN (the European Council for Nuclear Research) led to the invention of the World Wide Web—these are but some of the spin-offs from high-energy particle physics.

The applications of the technology and discoveries made in high-energy physics are legion, but it is not with this technological aim that the subject is pursued. The drive is curiosity; the desire to know what we are made of, where it came from, and why the laws of the universe are so finely balanced that we have evolved.

In this Very Short Introduction I hope to give you a sense of what we have found and some of the major questions that are now confronting us. I am indebted to Latha Menon at Oxford University Press for encouraging me to produce this new edition, and to Alfons Weber for his advice on some aspects of experimental particle physics.

List of illustrations

The publisher and the author apologize for any errors or omissions in
the above list. If contacted they will be pleased to rectify these at the
earliest opportunity.

Chapter 1
Journey to the centre
of the universe

Matter

On 4 July 2012, scientific epochs changed. That day at CERN, the particle physics laboratory in Geneva, discovery was announced of a new particle—the Higgs boson. This moment marked the 'end of the beginning' in the quest for matter's fundamental pieces and the laws that govern their behaviour. That the Higgs boson is the only single particle to have been named for one person itself hints at its unique role. That day marked the end of the 'pre-Higgs era', which had extended over 2,000 years, since the time when the ancient Greeks believed that everything is made from a few basic elements.

In the heat of the Big Bang, Higgs bosons filled the infant universe. If conditions allow, bosons can mingle in large numbers ('boson' is the descriptive name given to the species of particles that have that tendency, unlike 'fermions' which work more discretely). That is what happened to those Higgs bosons when the universe aged and cooled: they coalesced, and today slumber in a quiescent, ubiquitous field. To resurrect Higgs bosons from that field, it is necessary to recreate the heat energy that was prevalent in the early universe. That is what particle physics experiments at CERN can do, for brief moments, in a small region of space.

Discovery of the Higgs boson confirms that we are immersed in that all-pervading field. Its nature is yet a mystery, but its implications are not. Its presence gives rise to structure rather than a homogeneous goo. The interaction of fundamental particles with this field gives them mass, enabling structures like atoms and molecules, the templates of life.

Most of this book introduces what we know about the fundamental particles of matter and their role in building the universe at large during the pre-Higgs era. As to why they have these specific properties and not others, we do not know. Finding out is a goal for the post-Higgs era, which we shall meet at the end of the book.

The ancient Greeks' idea was basically correct; it was the details that were wrong. Their 'earth, air, fire, and water' are made of what today we know as the chemical elements. Pure water is made from two: hydrogen and oxygen. Air is largely made from nitrogen and oxygen with a dash of carbon and argon. The Earth's crust contains most of the 90 naturally occurring elements, primarily oxygen, silicon, and iron, mixed with carbon, phosphorus, and many others that you may never have heard of, such as ruthenium, holmium, and rhodium.

The abundance of the elements varies widely, and as a rough rule the ones that you think of first are among the most common, while the ones that you have never heard of are the rarest. Thus oxygen is the winner: with each breath you inhale 10 billion trillion atoms of it; so do the other 8 billion humans on the planet, plus innumerable animals, and there are plenty more oxygen atoms around doing other things. As you exhale these atoms are emitted, entrapped with carbon to make molecules of carbon dioxide, the fuel for trees and plants. The numbers are vast, and the names of oxygen and carbon are in everyone's lexicon. Contrast this with astatine or francium. Even if you have heard of them, you are unlikely to have come into contact with any, as it is estimated that there is less than an ounce of astatine in the Earth's crust, and as for francium it has even been claimed that at any instant there are at most 20 atoms of it around.

An atom is the smallest piece of an element that can exist and still be recognized as that element. Nearly all of these elements, such as the oxygen that you breathe and the carbon in your skin, were made in stars about 5 billion years ago, at around the time that the Earth was first forming. Hydrogen and helium are even older, most hydrogen having been made soon after the Big Bang, later to provide the fuel of the stars within which the other elements would be created.

A hundred years ago atoms were thought to be small, impenetrable objects, like miniature versions of billiard balls perhaps. Today we know that each atom has a rich labyrinth of inner structure. At its centre is a dense, compact nucleus, which accounts for all but a trifle of the atom's mass and carries positive electrical charge. In the outer regions of the atom there are tiny lightweight particles known as **electrons**. An electron has negative electric charge, and it is the mutual attraction of opposite charges that keeps these negatively charged electrons gyrating around the central positively charged nucleus.

The dot at the end of a sentence in printed text contains some 100 billion atoms of carbon. To see one of these with the naked eye, you would need to magnify the dot to be 100 metres across. While huge, this is still imaginable. But to see the atomic nucleus you would need that dot to be enlarged to 10,000 kilometres: as big as the Earth from pole to pole.

Between the compact central nucleus and the remote whirling electrons, atoms are mostly empty space. That is what many books assert, and while that may be true for the particles that make up an atom, it is only half the story. That space is filled with electric and magnetic force fields, so powerful that they would stop you in an instant if you tried to enter the atom. It is these forces that give solidity to matter, even while its atoms are supposedly 'empty'. As you read this, you are suspended an atom's breadth above the atoms in your chair due to these forces.

Powerful though these electric and magnetic forces are, they are trifling compared to yet stronger forces at work within the atomic nucleus. Disrupt the effects of these strong forces and you can release nuclear power; disrupt the electric and magnetic forces and you get the more ambient effects of chemistry and the biochemistry of life. These day-to-day familiar effects are due to the electrons in the outer reaches of atoms, far from the nucleus. Such electrons in neighbouring atoms may swap places, thereby helping to link the atoms together, making a molecule. It is the wanderings of these electrons that lead to chemistry, biology, and life. This book is not about those subjects, which deal with the collective behaviour of many atoms. By contrast, we want to journey into the atom and understand what is there.

Inside the atom

An electron appears to be truly fundamental; if it has any inner structure of its own, we have yet to discover it. The central nucleus, however, is built from further particles, known as **protons** and **neutrons**.

A proton is positively charged; the protons provide the total positive charge of the nucleus. The more protons there are in the nucleus, the greater is its charge, and, in turn, the more electrons can be held like satellites around it, to make an atom in which the positive and negative charges counterbalance, leaving the atom overall neutral. Thus it is that although intense electrical forces are at work deep within the atoms of our body, we are not much aware of them, nor are we ourselves electrically charged. The atom of the simplest element, hydrogen, normally consists of a single proton and a single electron. The number of protons in the nucleus is what differentiates one element from another. A cluster of 6 protons forms the nucleus of the carbon atom, iron has 26, and uranium 92.

Opposite charges attract, but like charges repel. So it is a wonder that protons, which are mutually repelling one another by this electrical force, manage to stay together in the confines of the nucleus. The reason is that when two protons touch, they can grip one another tightly by what is known as the strong force. This attractive force is much more powerful than the electrical repulsion, and so it is that the nuclei of our atoms do not spontaneously explode. However, you cannot put too many protons in close quarters; eventually the electrical disruption is too much. This is one reason why there is a heaviest naturally occurring element, uranium, with 92 protons in each nucleus. Pack more protons than this together and the nucleus cannot survive. Beyond uranium are highly radioactive elements such as plutonium, whose instability is infamous.

Atomic nuclei of all elements beyond hydrogen contain protons and also neutrons. The neutron is in effect an electrically neutral version of the proton. It has the same size and, to within a fraction of a percentage, the same mass as a proton. Neutrons grip one another with the same strength that protons do. Having no electrical charge, they feel no electrical disruption, unlike protons. As a result, neutrons add to the mass of a nucleus, and to the overall strong attractive force, and thereby help to stabilize the nucleus.

When neutrons are in this environment, such as when part of the nucleus of an iron atom, they may survive unchanged for billions of years. However, away from such a compact clustering, an isolated neutron is unstable. This is because a free neutron is a mere 1.4 parts in a thousand more massive than a proton, and by Einstein's equivalence of mass and energy, $E = mc^2$, contains correspondingly more energy. Nature seeks to find stability by minimizing this energy. It does so thanks to a feeble force, known as the weak force, one of whose effects is to destroy the neutron, converting it into a proton. This can even happen when too many neutrons are packed with protons in a nucleus. The effect of such a conversion here is to change the nucleus of one element into another. This transmutation of the elements is the seed of radioactivity and nuclear power.

Magnify a neutron or proton 1,000 times and you will discern that they too have a rich internal structure. Like a swarm of bees, which seen from afar appears as a dark spot whereas a close-up view shows the cloud buzzing with energy, so it is with the neutron or proton. On a low-powered image they appear like simple spots, but when viewed with a high-resolution microscope they are found to be clusters of smaller particles called **quarks**.

Let's take up the analogy of the full stop once more. We had to enlarge it to 100 metres to see an atom with the naked eye; to the diameter of the planet to see the nucleus. To reveal the quarks we would need to expand the dot out to the Moon, and then keep on going another 20 times further. In summary, the underlying structure of the atom is beyond real imagination.

We have at last reached the fundamental particles of matter as we currently know them. The electrons and the quarks are like the letters of Nature's alphabet, the basic pieces from which all can be constructed. If there is something more primary, like the dot and dash of Morse code, we do not know for certain what it is. There is speculation that if you could magnify an electron or a quark another billion billion times, you would discover the underlying Morse code to be like strings, which are vibrating in a universe that is revealed to have more dimensions than the three space and one time of which we are normally aware.

Whether this is the answer or not is for the future. I want to tell you something of how we came to know of the electron and the quarks, who they are, how they behave, and what questions confront us.

Forces

If the electrons and quarks are like the letters, then there are also analogues of the grammar: the rules that glue letters into words, sentences, and literature. For the universe, this glue is what we call the fundamental forces. There are four of them, of which gravity is the most familiar; gravity is the force that rules for bulk matter. Matter is held together by the electromagnetic force; it is this that holds electrons in atoms and links atoms to one another to make molecules and larger structures. Within and around the nucleus we find the other two forces: the strong and weak. The strong force glues the quarks into the small spheres that we call protons or neutrons; in turn these are held closely packed in the atomic nucleus. The weak force changes one variety of particle into another, such as in certain forms of radioactivity. It can change a proton into a neutron, or vice versa, leading to transmutation of the elements. In so doing it also liberates particles known as neutrinos. These are lightweight flighty neutral particles that respond only to the weak and gravitational forces. Millions of them are passing through you right now; they come from natural radioactivity in the rocks beneath your feet, but the majority have come from the Sun, having been produced in its central nuclear furnace, and even from the Big Bang itself.

For matter on Earth, and most of what we can see in the cosmos, this is the total cast of characters that you will need to meet. To make everything hereabouts requires the ingredients of electron and neutrino, and two varieties of quark, known as up and down, which seed the neutrons and protons of atomic nuclei. The four fundamental forces then act on these basic particles in selective ways, building up matter in bulk, and eventually you, me, the world about us, and most of the visible universe.

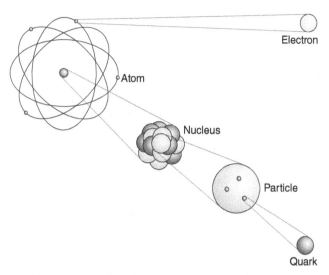

1. **Inside the atom. Atoms consist of electrons remotely encircling a massive central nucleus. A nucleus consists of protons and neutrons. Protons are positively charged; neutrons have no charge. Protons and neutrons in turn are made of yet smaller particles called quarks. To our best experiments, electrons and quarks appear to be basic particles with no deeper constituents.**

As a picture is said to be worth a thousand words, I summarize the story so far in the figures showing the inner structure of an atom and the forces of Nature.

How do we know this?

An important part of our story will be how we know these things. To sense the universe at all scales, from the vast extent of galaxies down to the unimaginably small distances within the atomic nucleus, requires that we expand our senses by using instruments. Telescopes enable us to look outwards and microscopes reveal what things are like at small distances. To look inside the atomic nucleus requires special types of microscope known as particle accelerators. By the use of electric fields, electrically charged particles such as electrons or protons are accelerated to within a fraction of the speed of light and then smashed into targets of matter or head on into one another. The results of such collisions can reveal the deep structure of matter. They not only show the quarks that seed the atomic nucleus, but have also revealed exotic forms of matter with whimsical names—strange, charm, bottom, and top—and seemingly heavier forms of the electron, known as the muon and tau. These play no obvious role in the matter that we normally find on Earth, and it is not completely understood why Nature uses them. Answering such questions is one of the challenges currently facing us.

Although these exotic forms are not prevalent today, it appears that they were abundant in the first moments after the Big Bang which heralded the start of our material universe. This insight has also come from the results of high-energy particle experiments, and a profound realization of what these experiments are doing. For 50 years the focus of high-energy particle physics was to reveal the deep inner structure of matter and to understand the exotic forms of matter that had unexpectedly shown up. In the last quarter of the 20th century a profound view of the universe emerged: that the material universe of today has emerged from a hot Big Bang, and that the collisions between subatomic particles are capable of recreating momentarily the conditions that were prevalent at that early epoch.

Thus today we view the collisions between high-energy particles as a means of studying the phenomena that ruled when the universe was newly born. We can study how matter was created and discover what varieties there were. From this we can construct the story of how the material universe has developed from that original hot cauldron to the cool conditions here on Earth today, where matter is made from electrons, without need for muons and taus, and where the seeds of atomic nuclei are just the up and down quarks, without need for strange or charming stuff.

Gravitational force

Electromagnetic force

Strong force

Weak force

e

p

n

ν

before

after

2. The forces of Nature. Gravity is attractive and controls the large-scale motions of galaxies, planets, and falling apples. Electric and magnetic forces hold electrons in the outer reaches of atoms. They can be attractive or repulsive, and tend to counterbalance in bulk

10

In very broad terms, this is the story of what has happened. The matter that was born in the hot Big Bang consisted of quarks and particles like the electron. The universe was filled with a ubiquitous essence, known as the 'Higgs field', but for which the vacuum of space would have been unstable. Interactions between the fundamental particles and the Higgs field gave particles mass. Why these masses have their specific values, we do not yet know. As concerns the quarks, the strange, charm, bottom, and top varieties are highly unstable, and they died out within a fraction of a second, the weak force converting them into their more stable progeny, the up and down varieties which survive within us today. A similar story took place for the electron and its heavier versions, the muon and tau. This latter pair are also unstable and died out, courtesy of the weak force, leaving the electron as survivor. In the process of these decays, lots of neutrinos and electromagnetic radiation were also produced, which continue to swarm throughout the universe some 14 billion years later.

The up and down quarks and the electrons were the survivors while the universe was still very young and hot. As it cooled, the quarks were stuck to one another, forming protons and neutrons. The mutual gravitational attraction among these particles gathered them into large clouds that were primaeval stars. As they bumped into one another in the heart of these stars, the protons and neutrons built up the seeds of heavier elements. Some stars became unstable and exploded, ejecting these atomic nuclei into space, where they trapped electrons to form atoms of matter as we know it. That is what we believe occurred some 5 billion years ago

matter, leaving gravity dominant at large distances. The strong force glues quarks to one another, forming neutrons, protons, and other particles. Its powerful attraction between protons and neutrons when they touch helps create the compact nucleus at the heart of atoms. The weak force can change one form of particle into another. This can cause transmutation of the elements, such as turning hydrogen into helium in the Sun.

when our solar system was forming; those atoms from a long-dead supernova are what make you and me today.

What we can now do in experiments is in effect reverse the process and observe matter change back into its original primaeval forms. Heat matter to a few thousand degrees and its atoms ionize—electrons are separated from the central nuclei. That is how it is inside the Sun. The Sun is a plasma, that is, gases of electrically charged electrons and protons swirling independently. At even higher temperatures, typical of the conditions that can be reached in relatively small high-energy accelerators, the nuclei are disrupted into their constituent protons and neutrons. At yet higher energies, these in turn 'melt' into a plasma of freely flowing quarks. At the extreme energies accessible by CERN's Large Hadron Collider, it has proved possible to excite the Higgs field, forming Higgs bosons. Thus we know we are immersed in this ubiquitous field, which gives fundamental particles their mass.

How this all happened, how we know, and what we've discovered are the themes of this Very Short Introduction.

Chapter 2
How big and small are big and small?

From quarks to quasars

The universe isn't the same everywhere—the Sun and stars are much hotter than the Earth and matter takes on different forms, but it is ultimately made of the same stuff. Atoms are very small; the cosmos is very big. How do they compare with everyday things?

Individual stars are huge, and visible to the naked eye over vast distances. This is in stark contrast to their basic components, the particles that eventually make up atoms. It would take about a billion atoms placed on top of one another to reach your head; it would take a similar number of people head to toe to give the diameter of the Sun. So this places the human measuring scale roughly in the middle between those of the Sun and an atom. The particles that make up atoms—the electrons that form the outer regions, and the quarks, which are the ultimate seeds of the central nucleus—are themselves a further factor of about a billion smaller than the atomic whole.

A fully grown human is a bit less than 2 metres tall. For much of what we will meet in this book, orders of magnitude are more important than precise values. So to set the scale I will take humans to be about 1 metre in 'order of magnitude' (this means we are much bigger than 1/10 metre, or 10^{-1} m, and correspondingly smaller than 10 m). Then, going to the large scales of astronomy, we have the radius of the Earth, some 10^7 m (that is, 1 followed by seven zeroes); that of the Sun is 10^9 m; our orbit around the Sun is 10^{11} m (or in more readable units, 100 million km). For later reference, note that the relative sizes of the Earth, Sun, and our orbit are factors of about 100.

Distances greater than this become increasingly hard to visualize, with large numbers of zeroes when expressed in metres, so a new unit is used: the light year. Light travels at 300,000 kilometres per second. This is fast but not infinite: it takes light a nanosecond, that is 10^{-9} s, to travel 30 cm, which is about the size of your foot. Modern computers operate on such timescales, and such microtimes will become central when we enter the world within the atom. For the moment, we are heading to the other extreme—the very large distances of the cosmos, and the long times that it takes for light to travel from remote galaxies to our eyes here.

It takes light 8 minutes to travel the 150 million km from the Sun; so we say the Sun is 8 light minutes away. It takes a year for light to travel 10^{16} m, and so this distance is referred to as a light year. Our Milky Way galaxy extends for 10^{21} m, or some 100,000 light years. Galaxies cluster together in groups, extending over 10 million light years. These clusters are themselves grouped into superclusters, about 100 million light years in extent (or 10^{24} m). The extent of the visible universe is some 10 billion light years, or 10^{26} m. These actual numbers are not too important, but notice how the universe is not homogeneous, and instead is clustered into distinct structures: superclusters, clusters of galaxies, and individual galaxies such as our own, with each being roughly 1/100 smaller than its predecessor. When we enter the microworld, we will once again experience such layers of structure, but on a much emptier scale; not 1/100, but more like 1/10,000.

Having made a voyage out into the large scales of space, let's now take the opposite direction into the microworld of atoms, and their internal structure. With our unaided naked eye, we can resolve individual pieces of dust, say, that are as small as a tenth to a hundredth of a millimetre: 10^{-4} to 10^{-5} m. This is at the upper end of the size of bacteria. Light is a form of electromagnetic wave, and the wavelength of visible light that we see as the rainbow spans 10^{-6} to 10^{-7} m. Atoms are a thousand times smaller than this: some 10^{-10} m. It is the fact that atoms are so much smaller than the wavelength of visible light that puts them beyond the reach of our normal vision.

Everything on Earth is made from atoms. Every element has its smallest piece, far too small to see by eye but real nonetheless, as special instruments can show.

To recap from Chapter 1: atoms are made of smaller particles. Electrons whirl in their remote reaches: at their heart is the compact massive atomic nucleus. The nucleus has a structure of its own, consisting of protons and neutrons, which in turn are made of yet smaller particles: the 'quarks'. Quarks and electrons are the seeds of matter as we find it on Earth.

Whereas the atom is typically 10^{-10} m across, its central nucleus measures only about 10^{-14} to 10^{-15} m. So beware the oft-quoted analogy that atoms are like miniature solar systems, with the 'planetary electrons' encircling the 'nuclear sun'. The real solar system has a factor 1/100 between our orbit and the size of the central Sun; the atom is far emptier, with 1/10,000 as the corresponding ratio between the extent of its central nucleus and the radius of the atom. And this emptiness continues. Individual protons and neutrons are about 10^{-15} m in diameter and are in turn made of yet smaller particles known as quarks. If quarks and electrons have any intrinsic size, it is too small for us to measure. All that we can say for sure is that it is no bigger than 10^{-18} m.

Particle Physics

10^{-18}	10^{-15}	10^{-12}	10^{-9}	10^{-6}	10^{-3}	1	10^{3}	10^{6}	10^{9}	10^{12}	10^{15}	10^{18}	10^{21}	10^{24}
	fm	pm	nm	μm	mm	m	km	Mm	Gm	Tm	Pm			

3. Comparisons with the human scale and beyond normal vision. In the small scale, 10^{-6} metres is known as 1 micron, 10^{-9} metres is 1 nanometre, and 10^{-15} metres is 1 fermi.

16

So here again we see that the relative size of quark to proton is some 1/10,000 (at most!). The same is true for the 'planetary' electron relative to the proton 'sun': 1/10,000 rather than the 'mere' 1/100 of the real solar system. So the world within the atom is incredibly empty.

To gain some sort of feel for this, imagine the longest hole that you are likely to find on a golf course, say 500 m. The relative length of this fairway to the size of the tiny hole into which you will eventually pot the ball is some 10,000:1 and hence similar to that of the radius of the hydrogen atom to its central nucleus, the proton.

Just as large distances become unwieldy when expressed in metres, so do the submicroscopic dimensions of atomic and nuclear structures. In the former case we introduced the light year, 10^{16} m; in the latter it is customary to use the angstrom, A, where 1 angstrom = 10^{-10} m (typically the size of a simple atom), and the fermi, fm, where 1 $fm = 10^{-15}\,m$. Thus angstroms are useful units to measure the sizes of atoms and molecules, while fermis are natural for nuclei and particles. (Ångström and Fermi were famous atomic and nuclear scientists of the 19th and 20th centuries, respectively.)

Our eyes see things on a human scale; our ancestors developed senses that would protect them from predators and had no need for eyes that could see galaxies that emit radio waves, or the atoms of our DNA. Today we can use instruments to extend our senses: telescopes that study the depths of space and microscopes to reveal bacteria and molecules. We have special 'microscopes' to reveal distances smaller than atoms: this is the role of high-energy particle accelerators. By such tools we can reveal Nature over a vast range of distance scales. How this is done for particles will be the theme of Chapters 5 and 6.

The universe in temperature and time

That is how things are now, but it hasn't always been that way. The universe, as we know it, began in a hot Big Bang where atoms could not survive. Today, about 14 billion years later, the universe at large is very cold, and atoms can survive. There are local hot spots, such as stars like our Sun, and matter there differs from that found here on our relatively cool Earth. We can even simulate the extreme conditions of the moments immediately following the Big Bang, in experiments performed at particle accelerators, and see how the basic seeds of matter originally must have emerged. However, although the forms that matter takes vary through space and time, the basic pieces are common. How matter appears in the cold (now), in the hot (such as in the Sun and stars), and in the ultra-hot (like the aftermath of the original Big Bang), is the theme of this section.

In macroscopic physics we keep our energy accounts in joules, or, in large-scale industries, mega- or terajoules. In atomic, nuclear, and particle physics, the energies involved are trifling in comparison. If an electron, which is electrically charged, is accelerated by the electric field of a 1-volt battery, it will gain an energy of 1.6×10^{-19} J. Even when rushing at near to the speed of light, as in accelerators at CERN in Geneva, the energy still only reaches the order of 10^{-8} J, one hundredth of a millionth of a joule. Such small numbers get messy, and so it is traditional to use a different measure, known as the 'electronvolt', or eV. We said above that when accelerated by the electric field of a 1-volt battery an electron will gain an energy of 1.6×10^{-19} J, and it is this that we define as 1 electronvolt.

Now the energies involved in subatomic physics become manageable. We call 10^3 eV a kilo-eV or keV; a million (mega), 10^6 eV, is 1 MeV; a billion (giga), 10^9 eV, is 1 GeV; and the latest experiments are entering the 'tera' or 10^{12} eV, 1 TeV, region.

Einstein's famous equation $E = mc^2$ tells us that energy can be exchanged for mass, and vice versa, the 'exchange rate' being c^2, the square of the velocity of light. The electron has a mass of 9×10^{-31} kg. Once again such numbers are messy, and so we use $E = mc^2$ to quantify mass and energy which gives about 0.5 MeV for the energy of a single electron at rest; we traditionally state its mass as 0.5 MeV/c^2. The mass of a proton in these units is 938 MeV/c^2, which is nearly 1 GeV/c^2.

Energy is profoundly linked to temperature also. If you have a vast number of particles bumping into one another, transferring energy from one to the next so that the whole is at some fixed temperature, the average energy of the individual particles can be expressed in eV (or keV and so on). Room temperature corresponds to about 1/40 eV, or 0.025 eV. Perhaps easier will be to use the measure of $1 \text{ eV} \Leftrightarrow 10^4 \text{ K}$ (where K refers to Kelvin, the absolute measure of temperature; absolute zero $0\text{K} = -273$ Celsius, and room temperature is about 300 K).

Fire a rocket upwards with enough energy and it can escape the gravitational pull of the Earth; give an electron in an atom enough energy and it can escape the electrical pull of the atomic nucleus. In many molecules, the electrons will be liberated by an energy of fractions of an eV; so room temperature can be sufficient to do this, which is the source of chemistry, biology, and life. Atoms of hydrogen will survive at energies below 1 eV, which in temperature terms is of the order of 10^4 K. Such temperatures do not occur normally on Earth (other than specific examples such as some industrial furnaces, carbon arc lights, and scientific apparatus), and so atoms are the norm here. However, in the centre of the Sun the temperature is some 10^7 K, or in energy terms 1 keV; atoms cannot survive such conditions.

At temperatures above 10^{10} K there is enough energy available that it can be converted into particles, such as electrons. An individual electron has a mass of 0.5 MeV/c^2, and so it requires 0.5 MeV of energy to 'congeal' into an electron. As we shall see later, this cannot happen spontaneously; an electron and its antimatter counterpart—the positron—must be created as a pair. So 1 MeV energy is needed for 'electron positron creation' to occur. Analogously, 2 GeV energy is needed to create a proton and its antiproton. Such energies are easy to generate in nuclear laboratories and particle accelerators today; they were the norm in the very early universe, and it was in those first moments that the basic particles of matter (and antimatter) were formed. The details of this will be given in Chapter 9, but some outline will be useful for orientation now.

The galaxies are observed to be rushing apart from one another such that the universe is expanding. From the rate of the expansion we can play the scenario back in time and deduce that about 14 billion years ago the universe would have been compacted in on itself. It is the explosive eruption from that dense state that we call the Big Bang. (It is not the primary purpose of this book to review the Big Bang; to learn more read Peter Coles' *Cosmology* in the Very Short Introduction series). In that original state, the universe would have been much hotter than it is now. The universe today is bathed in microwave radiation with a temperature of about 3 K. Combining this with the picture of the post-Big Bang expansion gives a measure of temperature of the universe as a function of time.

Within a billionth of a second of the original Big Bang, the temperature of the universe would have exceeded 10^{16} K, or in energy terms 1 TeV. The Higgs field filled and stabilized the vacuum. Some theorists ponder whether it helped drive the universe's initial inflationary expansion, but there is as yet no empirical proof of that. What we do know is that at such energies fundamental particles and antiparticles were created, including exotic forms no longer common today, and they gained mass through their interaction with the Higgs field. Most of these exotic

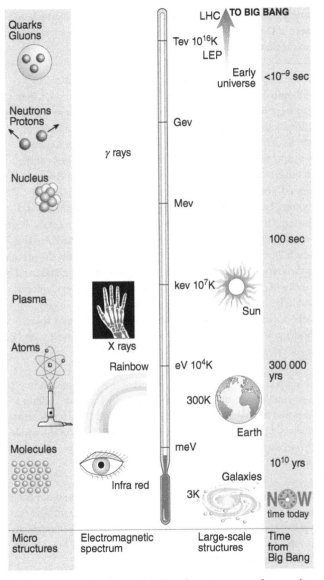

4. The correspondence between scales of temperature and energy in electronvolts (eV).

varieties died out almost immediately, producing radiation and more of the basic particles such as electrons and the surviving quarks that make up matter today.

As the universe aged, it cooled, at first very quickly. Within a millionth of a second quarks clustered together in threes, where they have remained ever since. So protons and neutrons were born. After about three minutes the temperature had fallen to about 10^{10} K, or in energy 1 MeV. This is 'cool' enough for protons and neutrons to stick together and build up the nuclear seeds of the (yet to be completed) atomic elements. A few light nuclei were formed, such as helium and traces of beryllium and boron. Protons, being stable and the simplest, were most common and clustered under gravity into spherical balls that we call stars. It was here that the nuclei of heavy elements would be cooked over the next billions of years. In Chapter 9 I shall describe how the protons in these stars bumped into one another, clustering together, and by a series of processes made the nuclear seeds of heavier elements: first helium, and eventually the heavier ones such as oxygen, carbon, and iron. When such stars explode and die, they spew these nuclear seeds out into the cosmos, which is where the carbon in your skin and the oxygen in our air originated.

The Sun is going through the first part of this story now. It has been converting protons into the nuclei of helium for 5 billion years and has used up about half of its fuel so far. The temperatures involved in its heart that do this are similar to those of the whole universe when it was a few minutes old. So the Sun is carrying on today what the universe did at large long ago.

Atoms cannot survive inside the depths of the Sun, and nor could they in the early universe. It was not until some 300,000 years had elapsed that the universe had cooled enough for these nuclei to entrap passing electrons and make atoms. That is how things are here on Earth today.

Chapter 3
How we learn what things are made of, and what we found

Energy and waves

Instruments such as microscopes and particle accelerators enable us to extend our vision beyond the rainbow of visible light and see into the subatomic microworld. This has revealed the inner structure of the atom—electrons, nuclear particles, and quarks.

To find out what something is made of you might (a) look at it; (b) heat it and see what happens; or (c) smash it by brute force. There is a common misconception that it is the latter that high-energy, or 'particle', physicists do. This is a term left over from the days when particle accelerators were known as 'atom smashers'. And indeed, historically that was what took place, but today the aims and methods are more sophisticated. We will come to the details later, but to start, let's focus on the three options just mentioned. Each of them shares a common feature: they all use energy.

In the case of heating, we have already seen how temperature and energy are correlated (10^4 K ~ 1 eV). Even in looking at things, energy will turn out to play a role.

You are seeing these words because light is shining on the page and then being transmitted to your eyes; the general idea here is that there is a source of radiation (the light), an object under investigation (the page), and a detector (your eye). Inside a full stop are millions of carbon atoms, and you will never be able to see the individual atoms, even with the most powerful magnifying glass. They are smaller than the wavelength of 'visible' light and so cannot be resolved in an ordinary magnifying glass or microscope.

Light is a form of electromagnetic radiation. Our eyes respond only to a very small part of the whole electromagnetic spectrum; but the whole of it can be accessed by special instruments. Visible light is the strongest radiation given out by the Sun, and humans have evolved eyes that register only this particular range. The whole spread of the electromagnetic spectrum is there, as we can illustrate by an analogy with sound. A single octave of sound involves a halving of the wavelength (or a doubling of the frequency) from one note (say the A at 440 Hz) to that of an octave above (the A at 880 Hz). Similarly for the rainbow: it is an 'octave' in the electromagnetic spectrum. As you go from red light to blue, the wavelength halves, the wavelength of blue light being half that of red (or equivalently, the frequency with which the electric and magnetic fields oscillate back and forth is twice as fast for blue light as red). The electromagnetic spectrum extends further in both directions. Beyond the blue horizon—where we find ultraviolet, X-rays, and gamma rays—the wavelengths are smaller than in the visible rainbow; by contrast, at longer wavelengths and in the opposite direction, beyond the red, we have infrared, microwaves, and radio waves.

We can sense the electromagnetic spectrum beyond the rainbow; our eyes cannot see infrared radiation, but the surface of our skin can feel it as heat. Modern infrared cameras can 'see' prowlers by the heat they give off. It is human genius that has made machines that can extend our vision across the entire electromagnetic range, thereby revealing deep truths about the nature of the atom.

Our inability to see atoms has to do with the fact that light acts like a wave and waves do not scatter easily from small objects.

To see a thing, the wavelength of the beam must be smaller than that thing is. Therefore, to see molecules or atoms needs illuminations whose wavelengths are similar to or smaller than them. Light waves, like those our eyes are sensitive to, have a wavelength of about 10^{-7} m (or put another way, 10,000 wavelengths would fit into a millimetre). This is still 1000 times bigger than the size of an atom. To gain a feeling for how big a task this is, imagine the world scaled up 10 million times. A single wavelength of light, magnified 10 million times, would be bigger than a human, whereas an atom on this scale would extend only 1 millimetre, far too little to disturb the long blue wave. To have any chance of seeing molecules and atoms we need light with wavelengths much shorter than these. We have to go far beyond the blue horizon to wavelengths in the X-ray region and beyond.

X-rays are light with such short wavelengths that they can be scattered by regular structures on the molecular scale, such as are found in crystals. The wavelength of X-rays is larger than the size of individual atoms, so the atoms are still invisible. However, the distance between adjacent planes in the regular matrix within crystals is similar to the X-ray wavelength, and so X-rays begin to discern the relative position of things within crystals. This is known as 'X-ray crystallography'.

An analogy can be made if one thinks for a moment of water waves rather than electromagnetic ones. Drop a stone into still water and ripples spread out. If you were shown an image of these circular patterns, you could deduce where the stone had been. A collection of synchronized stones dropped in would create a more complicated pattern of waves, with peaks and troughs as they meet and interfere. From the resulting pattern you could deduce, with some difficulty admittedly, where the stones had entered. X-ray crystallography involves detecting multiple scattered waves from the regular layers in the crystal and then decoding the pattern to deduce the crystalline structure. In this way, the shape and form of very complicated molecules, such as DNA, have been deduced.

To resolve the individual atoms we need even shorter wavelengths, and we can do this by using not just light, but also beams of particles such as electrons. These have special advantages in that they have electric charge and so can be manipulated, accelerated by electric fields, and thereby given large amounts of energy. This enables us to probe ever shorter distances, but to understand why we need to make a brief diversion to see how energy and wavelength are related.

One of the great discoveries in the quantum theory was that particles can have wavelike character, and conversely that waves can act like staccato bundles of particles, known as 'quanta'. Thus an electromagnetic wave acts like a burst of quanta—photons. The energy of any individual photon is proportional to the frequency (v) of the oscillating electric and magnetic fields of the wave. This is expressed in the form

$$E = hv$$

where the constant of proportion, h, is Planck's constant.

The length of a wave (λ), and the frequency with which peaks pass a given point, are related to its speed, c, by $v = c / \lambda$. So we can relate energy and wavelength

$$E = \frac{hc}{\lambda}$$

and the proportionality constant $hc \sim 10^{-6}$ eV m. This enables us to relate energy and wavelength by the approximate rule of thumb: '1 eV corresponds to 10^{-6} m', and so on.

You can compare with the relation between energy and temperature in Chapter 2, and see how temperature and wavelength are related. This illustrates how bodies at different temperatures will tend to radiate at different wavelengths: the

ENERGY	Wavelength (m)
1 eV	10^{-6}
1 keV	10^{-9}
1 MeV	10^{-12}
1 GeV	10^{-15}
1 TeV	10^{-18}

5. Energy and approximate wavelengths.

hotter the body, the shorter the wavelength. Thus, for example, as a current flows through a wire filament and warms it, it will at first emit heat in the form of infrared radiation—and as it gets hotter, 1,000 degrees or so, it will begin to emit visible light and illuminate the room. Hot gases in the vicinity of the Sun can emit X-rays; some extremely hot stars emit gamma rays.

To probe deep within atoms we need a source of very short wavelength. As we cannot make gamma-emitting stars in the laboratory, the technique is to use the basic particles themselves, such as electrons and protons, and speed them in electric fields. The higher their speed, the greater their energy and momentum and the shorter their associated wavelength. So beams of high-energy particles can resolve things as small as atoms. We can look at as small a distance as we like; all we have to do is to speed the particles up, give them more and more energy to get to ever smaller wavelengths. To resolve distances on the scale of the atomic nucleus, 10^{-15} m, requires energies of the order of GeV. This is the energy scale of what we call high-energy physics. Indeed, when that field began in earnest in the early to middle of the 20th century, GeV energies were at the boundaries of what was technically available. By the end of the 20th century, energies of several hundred GeV were the norm, and we are now entering the realm of TeV scale energies, probing matter at distances smaller than 10^{-18} m. So when we say that electrons and quarks have no deeper structure, we can only really say 'at least on scales of 10^{-18} m'. It is possible that there are deeper layers, on distances smaller than these, but which are beyond our present ability to resolve in experiment. So although I shall throughout this book speak as if these entities are the ultimate pieces, always bear in mind that caveat: we only know how Nature operates at distances larger than about 10^{-18} m.

Accelerating particles

The idea of accelerators will be described in Chapter 5, but for the moment let's reflect a moment on what is required. To accelerate particles to high energy requires lots of space. Technology in the mid- to late -20th century could accelerate electrons, say, at a rate corresponding to each electron in the beam gaining some tens of MeV energy per metre travelled. Hence the Stanford Linear Accelerator Center in California (SLAC) contained a 3-km-long accelerator which produced beams of electrons at up to 50 GeV, while in the 1990s at CERN in Geneva, electrons were guided around a circle of 27 km in length, achieving energies of some 100 GeV. Protons, being more massive, pack a bigger punch, but still require large accelerators to achieve their goals. Today, that 27-km tunnel at CERN houses the Large Hadron Collider, which accelerates protons and heavy nuclei to above 1 TeV—1,000 GeV. Ultimately it is the quantum relation between short distances, the consequent short wavelengths needed to probe them, and the high energies of the beams that creates this apparent paradox of needing ever bigger machines to probe the most minute distances.

These were the early aims of those experiments to probe the heart of the atomic nucleus by hitting it with beams of high-energy particles. The energy of the particles in the beam is vast (on the scale of the energy contained within a single nucleus, holding the nucleus together), and as a result the beam tends to smash the atom and its particles apart into pieces, spawning new particles in the process. This is the reason for the old-fashioned name of 'atom smashers'. Today we do much more than this, and the name is defunct.

The electron and proton

The electrically charged particles that build up atoms are the electron and proton. An atom of the simplest element, hydrogen, consists normally of a single electron (negatively charged) and a proton with the same amount of charge, but positive. Thus, although an atom can be electrically neutral overall (as is the case with most bulk matter that we are familiar with), it contains negative and positive charges within. It is these charges, and the consequent electric and magnetic forces that they feel, which bind atoms into molecules and bulk matter. We will deal with the forces of Nature in Chapter 7. Here we will focus on these basic electrically charged particles and how they have been used as tools to probe atomic and nuclear structure.

Electron beams were being used in the 19th century, though no one then knew what they were. When electric currents were passed through gases at extremely low pressures, a pencil-thin beam would be seen. Such beams became known as 'cathode rays' and, we now know, consist of electrons. The most familiar example of this apparatus is a classic television, where the cathode is the hot filament at the rear from which the beams of electrons emerge and hit the screen.

It was a big surprise in the 19th century when it was discovered that the rays would pass through solid matter almost as if nothing was in their way. This was a paradox: matter that is solid to the touch is transparent on the atomic scale. Phillipp Lenard, who discovered this, remarked that 'the space occupied by a cubic metre of solid platinum is as empty as the space of stars beyond the earth'. Atoms may be mostly empty space, but something defines them, giving mass to things. That there is more than simply space became clear with the work of Ernest Rutherford in the early years of the 20th century. This came about after the discoveries of the electron and of radioactivity, which provided the essential tools with which atomic structure could be exposed.

The electron was discovered and identified as a fundamental constituent of the atomic elements by J. J. Thomson in 1897. Negatively charged, electrons have been inside atoms as long as the Earth has been here. They are easy to extract, temperatures of a few thousand degrees will do. Electric fields will accelerate them, giving them energy and thereby enabling beams of high-energy electrons to probe small-scale structures.

There are other atomic bullets. The proton has positive electric charge, in magnitude the same as the electron's negative, but in mass the proton wins out immensely, being nearly 2,000 times as massive. Protons have become a choice beam for subatomic investigations, but initially it was another electrically charged entity that proved seminal. This was the alpha particle.

Today we know that this is the nucleus of a helium atom: a compact cluster of two protons and two neutrons, and as such, positively charged and some four times as massive as a single atom of hydrogen. The reason that this came to prominence is that the nuclei of many heavy elements are radioactive, spontaneously emitting alpha particles and thereby providing freely a source of electrically charged probes. Heavy nuclei consist of large numbers of protons and neutrons tightly packed, and the phenomenon of alpha radioactivity occurs as a heavy nucleus gains stability by spontaneously ejecting a tight bunch of two protons and two neutrons. The details of this need not concern us here, suffice to accept that it occurs, that the 'alpha' particle emerges with kinetic energy and can smash into the atoms of surrounding material. It was by such means that Ernest Rutherford and his assistants Geiger and Marsden first discovered

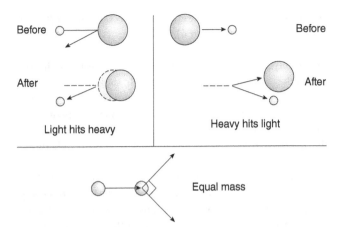

6. Result of heavy and light objects hitting light and heavy targets, respectively.

the existence of the atomic nucleus. (For more about nuclear physics, see *Nuclear Physics, A Very Short Introduction*.)

When alpha particles encountered atoms, the alphas were sometimes scattered violently, even on occasion being turned back in their tracks. This is what happens if the positive charge of a heavy element, such as gold, is concentrated in a compact central mass. The positively charged alphas were being repelled by the positively charged atomic nucleus; and as a light object, such as a tennis ball, can recoil from a heavy one, such as a football, so did the alphas recoil from the massive nucleus of the gold atom.

Alphas are much lighter than the nuclei of gold but heavier than a proton, the nucleus of the hydrogen atom. So if alpha particles are fired at hydrogen, one would have a situation akin to the football hitting the lightweight tennis ball. In such a case, the football will tend to carry on in its flightpath, knocking the tennis ball forwards in the same general direction. So when the relatively massive alphas hit the protons of hydrogen, it is these protons that are ejected forwards. These were detected by the trails they left in cloud chambers (see Chapter 6).

By such experiments in the early years of the 20th century, the basic idea of the nuclear atom was established. To summarize: the way that the alpha particles scattered from atoms helped to establish the picture of the atom that we have known ever since: the positive charge lives in a compact bulky centre—the atomic nucleus—while the negatives are electrons whirling remotely on the periphery.

Naturally occurring alpha particles don't pack much punch. They are ejected from heavy nuclei with only a few MeV kinetic energy, or equivalently a few MeV/c momentum, and as such are able to resolve structures on distance scales larger than about 10^{-12} m. Now, such sizes are smaller than those of atoms, which makes such alphas so useful, but they are still much larger than the 10^{-14} m extent of even a large nucleus, such as that of a gold atom, let alone the 10^{-15} m size of the individual protons and neutrons that combine to make that nucleus. So although alphas were fine for discovering the existence of the atomic nucleus, to see inside such nuclei would require beams with more energy.

With this as the aim, we have here the beginnings of modern high-energy physics. It was in 1932 that the first accelerator of electrically charged particles was built by Cockroft and Walton, and a detailed picture of nuclear structure, and of the particles that build it, began to emerge. One can use beams of atomic nuclei, but while these were truly 'atom (or rather nuclear) smashers', and helped to determine the pattern of nuclear isotopes (forms of the same element that contain equal numbers of protons but different numbers of neutrons) and their details, the clearest information on their basic constituents came with the simplest beams. A nucleus of carbon contains typically six protons and a similar number of neutrons. As such there is a lot of debris when it hits another nucleus, some coming from the carbon beam itself as well as that from the target. This makes a clear interpretation difficult. It is far cleaner to use a beam of just protons; this was, and remains today, one of the main ways of probing the nucleus, and distances down to 10^{-19} m.

Protons, which carry the positive charge, have been favourites for over 50 years, as they pack a big punch. However, electrons have some special advantages, and much of our present knowledge about the structure of atomic nuclei, and even the protons and neutrons from which they are made, is the result of experiments using electron beams.

Radioactivity in the form of beta decay emits electrons—the 'beta' radiation—which could be used to probe atomic structure. However, such electrons have energies of only a few MeV, as was the case for alpha particles, and so suffer the same limitations: they allow us to see a nucleus like the alpha can, but they cannot resolve the inner structure of the nucleus. The key to progress was to ionize atoms, liberating one or more of their electrons, and then accelerate the accumulated electron beam by means of electric fields. By the 1950s in Stanford, California, beams with energies of 100 MeV to 1 GeV per electron began to resolve distances approaching 10^{-15} m. The electrons scattered from the protons, and neutrons began to reveal evidence of a deeper layer of structure within those nuclear particles. Such experiments showed that the neutron, though electrically neutral overall, has magnetic effects and other features suggesting there is charge within it, positive and negative counterbalancing somehow, as had been the case in atoms. Protons too were found to have a finite size, extending over a distance of order 10^{-15} m. Once it was established that protons are not point particles, the question arose as to how the charge of a proton is distributed within its size. Such questions are reminiscent of what had happened years before in the case of atoms, and the answers came by similar experiments. In the case of the atom, its hard nuclear core was revealed by the scattering of alpha particles; in the case of the proton, it would be beams of high-energy electrons that would give the answer.

It was the 3-km-long linear accelerator of electrons at Stanford that in 1968 took the first clean look inside the atomic nucleus and discovered that what we know as protons and neutrons are actually little spheres of swarming 'quarks'.

At energies above 10 GeV, electrons can probe distances of 10^{-16} m, some ten times smaller than the proton as a whole. When they encountered the proton, the electrons were found to be scattered violently. This was analogous to what had happened 50 years earlier with the atom; where the violent scattering of relatively low-energy alpha particles had shown that the atom has a hard centre of charge, its nucleus, the unexpected violent scattering of high-energy electron beams showed that a proton's charge is concentrated on 'pointlike' objects—the quarks (pointlike in the sense that we are not able to discern whether they have any substructure of their own). In the best experiments that we can do today, electrons and quarks appear to be the basic constituents of matter in bulk.

Chapter 4
The heart of the matter

At the heart of ordinary matter, we find up and down quarks and the electron, while radioactivity reveals the ghostly neutrino. In this chapter we will see the roles they play and how their masses and other properties are critical for making life, the universe, but not everything. Cosmic rays have revealed evidence of extra-terrestrial forms of matter that do not occur naturally here on Earth. Meanwhile, the Sun and stars produce vast numbers of neutrinos, which is inspiring a new science: neutrino astronomy.

We have described how the discovery a century ago of atomic structure and of the proton came about as a result of scattering beams of high-energy particles from them. However, in the case of both atoms and protons, the first hints that they had a substructure came earlier, from the discovery of spectra.

The first clue to the existence of electrons within atoms was the discovery that atomic elements emit light with discrete wavelengths manifested, for example, as discrete colours rather than the full spread of the rainbow, so-called spectral lines. We now know that quantum mechanics restricts the states of motion of electrons within atoms to a discrete set, each member of which has a specific magnitude of energy. The configuration where an atom has the lowest total energy is known as the 'ground state'; all other configurations have larger energies and are known as excited states. Atomic spectra are due to light radiated or absorbed when their electrons jump between excited states, or between an excited state and the ground state. Energy is conserved overall; the difference in energy of the two atomic states is equal to the energy of the photon that has been emitted or absorbed in the process. It was the spectra of these photons that revealed the differences in these energy levels of the atom, and from the rich set of such data a picture of the energy levels could be deduced. Subsequently the development of quantum mechanics explained how the pattern of energy levels emerges: it is determined by the nature of the electric and magnetic forces binding the electrons around the central nucleus—in particular for the simplest atom, hydrogen, being intimately linked to the fact that the strength of the electrical force between its electron and proton falls as the square of the distance between them.

An analogous set of circumstances occurred in the case of the proton. When experiments with the first 'atom smashers' took place in the 1950s to 1960s, many short-lived heavier siblings of the proton and neutron, known as 'resonances', were discovered. A panoply of states emerged, and with hindsight it is obvious, though it was not so at the time, that here was evidence that the proton and neutron are composite systems made, as we now know, from quarks. It is the motion of these quarks that gives size to the proton and neutron, analogous to the way that the motion of electrons determines the size of atoms. It is the quarks that provide the electric charge and the magnetic properties of a proton or neutron. Although the electric charges of the quarks that form a neutron add up to zero, their individual magnetism does not cancel out, which leads to the magnetic moment of the neutron. It is when the quarks are in the state of lowest energy that the configurations that we call proton and neutron arise: excite one or more quarks to a higher energy level in the potential that binds them and one forms a short-lived resonance with a correspondingly larger rest-energy, or mass. Thus the spectroscopy of short-lived resonance states is due to the excitation of the constituent quarks.

This far is akin to the case of atoms. However, there are some important differences. When more and more energy is given to the electrons in an atom, they are raised to ever higher energy levels, until eventually they are ejected from the atom; in this type of scenario we say that the atom is 'ionized'. In Chapter 2 we saw how a temperature of 10^4 K provides enough energy to ionize atoms, as happens in the Sun. In the case of the proton, as it is hit with ever higher energies, its quarks are elevated to higher levels, and short-lived resonances are seen. This energy is rapidly released, by emitting photons or, as we shall see, other particles, and the resonance state decays back to the proton or neutron once more. No one has ever ionized a proton and liberated one of its constituent quarks in isolation: the quarks appear to be permanently confined within a region of about 10^{-15} m—the 'size' of the proton. Apart from this, which is a consequence of the nature of the forces between the quarks, the story is qualitatively similar to that of electrons within atoms. The excited levels are short lived, and release excess energy, typically by radiating energy in the form of gamma-ray photons, and fall back to the ground state (proton or neutron). Conversely one can excite these resonance states by scattering electrons from protons and neutrons.

The final piece to the analogy came around 1970. Beams of electrons, which had been accelerated to energies of over 20 GeV, were scattered from protons at Stanford in California. Similar to what had occurred for Rutherford half a century earlier, the electrons were observed to scatter through large angles. This was a direct consequence of the electrons colliding with quarks, the pointlike fundamental particles that comprise the proton.

These experiments have been extended to higher energies, most notably at the HERA accelerator in Hamburg, Germany. The resulting high-resolution images of the proton have given fundamental insights into the nature of the forces binding the quarks to one another. This has given rise to the theory of quarks known as quantum chromodynamics, of which we shall learn more in Chapter 7. Its ability to describe the interactions of quarks and gluons at distance scales below 10^{-16} m has passed every experimental test.

Quarks with flavour

Three quarks clustered together are sufficient to make a proton or a neutron. There are two different varieties (or 'flavours') of quark needed to make a proton and neutron, known as the **up** and **down** (traditionally summarized by their first letters, u and d, respectively). Two ups and one down make a proton; two downs and one up make a neutron.

The quarks are electrically charged. An up quark carries a fraction 2/3 of the (positive) charge of a proton, while a down quark carries a fraction –1/3 (that is, negative). Thus as the total electric charge of a collection is the sum of the individual pieces, we have for the charge of a proton $p(uud) = 2/3 + 2/3 - 1/3 = +1$, and of a neutron $n(ddu) = -1/3 - 1/3 + 2/3 = 0$.

Particles have an intrinsic angular momentum, or 'spin'. The amount of spin is measured in units of Planck's quantum, h divided by 2π; as this combination occurs throughout atomic and particle physics it is denoted by the symbol $\hbar$. The proton,

quark	Up	Down
Charge	+2/3	–1/3
Mc^2(MeV)	$\simeq 2$	$\simeq 5$
Spin	1/2	1/2

7. **Properties of up and down quarks.**

One quark can point its spin axis up or down q↑ or q↓

Two quarks with net spin 1 ↑↑ or 0 ↑↓

Three quarks with net spin 3/2 ↑↑↑ or 1/2 ↑↓↑

Examples $\left(\begin{array}{c} \text{uud} \\ \uparrow\uparrow\uparrow \end{array}\right) = \Delta^+$ $\left(\begin{array}{c} \text{uuu} \\ \uparrow\uparrow\uparrow \end{array}\right) = \Delta^{++}$ with spin 3/2

$\left(\begin{array}{c} \text{uud} \\ \uparrow\downarrow\uparrow \end{array}\right)$ or $\left(\begin{array}{c} \text{uud} \\ \uparrow\uparrow\downarrow \end{array}\right)$ with spin 1/2 = proton

8. Quark spins and how they combine.

neutron, and the quarks each have an amount $\hbar/2$, or in the usual shorthand, 'spin 1/2'.

Spins add or subtract so long as the total is not negative. So combining two particles each having spin 1/2 gives either 0 or 1. Adding three together gives a total of either 1/2 or 3/2. The proton and neutron have spin 1/2 resulting from the three quarks having coupled their individual spins to the former possibility. When the quarks combine to a total of 3/2, they have slightly greater total energy, and this forms the short-lived particles known as the 'Δ resonances', which have some 30 per cent more mass than do the proton or neutron, and they survive less than 10^{-23} s before decaying back to the more stable neutron or proton. (10^{-23} s is a time too short to imagine, but roughly it is similar to the time that it takes light to travel across a single atomic nucleus.) Rules of quantum theory (the 'Pauli exclusion principle') allow only certain correlations to occur among the spins and flavours of the quarks, and it is this that ultimately forbids three 'identical' up quarks (or three down) to combine into a net spin 1/2; thus there are no siblings of the proton and neutron with charge +2 or –1 made respectively of uuu and ddd. By contrast, when the three quarks have coupled their spins to a total of 3/2, three identical 'flavours' of quark are allowed to cluster together. Thus there exist examples such as the $\Delta^{++}(uuu)$ and $\Delta^-(ddd)$ (with superscripts denoting their electric charges). The full details of how these correlations emerge involve properties of the quarks that govern the strong interquark forces (see Chapter 7), but go beyond the scope of this short introduction.

The individual quarks have masses that are about ten times larger than that of an electron. As protons and neutrons have similar masses to one another, and nearly 2,000 times greater than that of an electron, there are two questions to face. One is: how do the proton and neutron get such large masses; the other is: are the masses of these quarks perhaps to be regarded as similar to that of the electron, hinting at some deeper unity among the fundamental constituents of matter?

Quarks grip one another so tightly that they are forever imprisoned in groups, such as the threesome that forms the entity that we call the proton. No quark has ever been isolated from such a family; their universe extends only for the 10^{-15} m that is the extent of the proton's size, and it is this confinement within the 10^{-15} m 'femtouniverse' that we call the proton that gives them collectively an energy of ~ 938 MeV, which is the mass of the proton. We saw how length and energy are related, and that distances of the order of 10^{-15} m correspond to an energy of around 1 GeV. The precise correspondence of relevance here involves factors of 2 and π, which go beyond this Very Short Introduction, with the result that an up or down quark, which were it free would have a mass of only a few MeV, when restricted to a femtouniverse of 10^{-15} m has an energy of some 200–300 MeV. The quarks are interacting strongly with one another (which must be so as they do not escape!), and the full details of how the mass of the proton turns out to be precisely 938.4 MeV is beyond our ability to derive from theory at present.

The down quark is a few MeV more massive than the up quark. We don't know why this is (indeed, we don't know why these fundamental particles, along with the electron, have the masses they do), but this does explain why the neutron is slightly more massive than the proton. A trio as in *uud* (proton) and *ddu* (neutron) each have mass of around 1 GeV due to their common entrapment in a 10^{-15} m region. There will be differences at the order of an MeV as a result of two features: (i) the neutron has an extra down quark at the expense of an up quark relative to the proton, and the greater mass of this down quark gives the total in forming the neutron a greater mass than the corresponding trio for a proton; (ii) the electrostatic forces among two ups and a down (as in a proton) will differ from those between two downs and an up (as in a neutron). These also contribute to the total energy at the MeV scale. So the mass difference between a neutron and proton (experimentally 1.3 MeV) is due to the electrostatic forces between their constituent quarks and the greater intrinsic mass of a down quark relative to the up quark.

Up and down are siblings in the quark family. The electron is not made of quarks, and as far as we know is itself fundamental, like the quarks. As such it belongs to a different family, known as leptons. As up and down quarks are paired, with a difference of one unit between their respective electric charges (in the sense that $+2/3 - (-1/3) = 1$), so does the electron have a sibling whose electric charge differs from the electron's by one unit. This entity, with no electric charge, is known as the **neutrino**.

Neutrinos are produced in radioactive decays of many atomic nuclei. In these processes they appear along with their sibling, the electron. For example, so long as it is not trapped in a nucleus, a neutron turns into a proton by emitting an electron and a neutrino in the process. This is called beta decay, where the instability of the neutron is due to it having a slightly greater mass than does a proton. Nature seeks the state of lowest energy, which translates in this case to the state of lowest mass. It is the small excess mass of a neutron that makes it (slightly) unstable when left in isolation. If you had a large sample of neutrons, each of them free of the others, then after about 10 minutes, half will have decayed by beta radioactivity. If we denote the neutron and proton

9. Beta decay of a neutron.

by the symbols n, p, and the electron and neutrino by e^-, ν, then beta decay of the neutron is summarized by the expression

$$n \rightarrow p + e^- + \nu$$

The neutron has no electrical charge overall; this is preserved in the beta decay as the proton has one unit positive, counterbalancing the negative electron. The proton, being the lightest state made of three quarks, is stable (or, at least, if protons are unstable, their mean lifetime is greater than 10^{32} *years*!) Technically an **anti-**neutrino is produced: see later section, 'Antiparticles'.

The neutrino

If we imagined removing the electric charge from the electron and also (almost) all of its mass, the resulting electrically neutral entity would be in effect a neutrino. Along with no electric charge, the neutrino has almost no mass and goes through almost everything. Oblivious to the normal electrical forces that act within bulk matter, neutrinos are hard to detect. It is figuratively the most nugatory of the particles.

The neutrino is the first 'fossil' relic of the Big Bang, and a messenger from the earliest processes in the universe. Neutrinos determine how fast the universe is expanding, and may determine its ultimate destiny. In stars like the Sun, they are essential in helping to cook the heavy elements that are necessary for life. The Sun is powered by the fusion of protons near its centre bumping into one another, joining and building up the nuclei of helium.

In doing so some protons turn into neutrons by a form of beta radioactivity, and neutrinos are emitted as this happens. The effect is enormous: neutrinos are produced in the Sun at a rate of 2×10^{38} each second. That's 2 followed by 38 zeroes; I cannot even imagine how to give an idea of how huge that number is—it's like the relative size of the whole universe to the size of a single atom. These neutrinos fly out into space and many hit Earth. About 400 billion neutrinos from the Sun pass through each one of us each second.

Natural radioactivity of the elements in the ground, such as uranium, also liberate neutrinos: about 50 billion of them hit us each second. So the Sun is indeed putting out a lot: eight times as many arrive from the Sun each second after spreading out over 100 million km of space than come from beneath our feet here at home. And we ourselves are radioactive (mainly from the decays of potassium in our bones) and emit some 400 neutrinos a second.

All in all, neutrinos are the commonest particles of all. There are even more of them flying around the cosmos than there are photons, the basic particles of light.

Neutrinos from the Sun fly through matter almost unchecked, so as many fly up through our beds at night as shine down on our heads by day. One of these neutrinos could fly through a light year of lead without hitting anything. This property of the neutrino is frequently mentioned in popular articles, and begs an obvious question: how do we detect them? Two things come to our aid.

The first is to use very intense sources of neutrinos so that the lottery of chance means that one or two will bump into atoms in some detector and be recorded. Although a single neutrino might only interact once in a blue moon (or a light year), the Sun is putting out so many that chance provides assistance. You or I have almost no chance of winning the lottery, but enough people enter that someone does. With enough neutrinos shining down on us, and a large amount of detector material, a few will hit atoms in the detector and reveal themselves. As a result, it has been possible to detect neutrinos coming from the Sun, from supernova SN1987A, and also ultra-high energy neutrinos from the cosmos. A new science, called neutrino astronomy, is now beginning.

The second property that comes to our aid is that their 'shyness' is only true for neutrinos with low energies, such as those emitted by the Sun. By contrast, neutrinos with high energy (as produced in some cosmic processes or in high-energy particle accelerators) have much greater propensity to reveal themselves. So it is in high-energy accelerators that we have produced neutrinos and studied them in detail. And it is here that we are getting our first hints that neutrinos do have a small, but non-zero, mass. These data also reveal that neutrinos occur in three distinct varieties, or 'flavours' (Chapter 8). For now, by 'neutrino' I shall mean the most familiar one partnered with the electron.

Antiparticles

The quarks and the electron are the basic seeds of atoms, and of matter as we know it. But they are not the full story; they also occur in a sort of mirror image form, known as antiparticles, the seeds of antimatter. Every variety of particle has as its 'anti' version; an entity with the same mass, spin, size, and amount of electric charge as itself, but with the sign of that charge reversed. So, for example, the negatively charged electron has as its anti-electron a positively charged entity, which is known as the **positron**, not to be confused with the proton. A proton is nearly 2,000 times more massive than a positron, and has its own anti-version, the antiproton, which has negative charge.

The forces that enable an electron and proton to combine to form an atom of hydrogen also enable a positron and antiproton to form an atom of antihydrogen. At CERN it is possible to create and store atoms of antihydrogen. This enables the properties of antihydrogen to be compared with those of hydrogen. Finding any difference between these two forms of substance would shake the foundations of particle physics. As of 2023, no difference has been found.

We can summarize the charges of the basic particles and antiparticles that we have met so far in the table in Figure 10.

As a proton is made of uud, so is the antiproton made of the corresponding antiquarks, $\overline{uud}$. It is traditional to denote an antiparticle by the symbol of the corresponding particle but with a line over the top. This is so unless the charge is specified, in which case the antiparticle is of the opposite charge (for example the positron, which is uniformly denoted as e^+ for historical reasons). Similarly for the neutron ddu, the antineutron is made of $\overline{dd\,u}$. So although a neutron and antineutron have the same electrical charge, their inner structure distinguishes them. A neutrino and antineutrino also have the same charge, but their distinguishing property is more subtle. When neutrinos interact with a particle of matter, a neutron say, they will turn into electrons and the neutron is converted into a proton, thereby preserving overall the electric charge:

$$\nu + n \rightarrow e^- + p$$

PARTICLE	CHARGE	ANTIPARTICLE	CHARGE
electron e^-	−1	positron e^+	+1
neutrino ν	0	antineutrino $\bar{\nu}$	0
up quark u	+2/3	antiup $\bar{u}$	−2/3
down quark d	−1/3	antidown $\bar{d}$	+1/3

10. **Fundamental particles of matter and their antiparticles.**

In this sense we see that the neutrino has an affinity for the electron. An antineutrino has an analogous affinity for the positron. The conservation of electric charge then prevents an antineutrino interacting with a neutron to make an analogue of the above, but if it hits a proton, it can reveal itself.

$$\bar{\nu} + p \rightarrow e^+ + n$$

We have seen how three quarks unite to make particles such as the proton and neutron (generically these three-quark composites are known as **baryons**). The clusters of three antiquarks are then known collectively as antibaryons. It is possible to cluster quarks and antiquarks; one of each is sufficient. So if we use q to denote either of u or d, and $\bar{q}$ to denote the antiquarks, it is possible to make four combinations of clusters $q\bar{q}$. As a three-quark cluster is called a baryon, so this combination of quark and antiquark is known as a **meson**. As was the case for the proton and neutron, there are higher-energy 'resonant' states for these mesons too.

One of the most famous properties of antimatter is that when it meets with matter, the two mutually annihilate in a flash of radiation, such as photons of light. It is no surprise then that mesons do not live very long. A quark and an antiquark, restricted to the femtouniverse of 10^{-15} m, mutually annihilate within a billionth of a second, or less. Even so, such ephemeral mesons play a role in building our universe. The most familiar, and the lightest, configuration are the **pions**, such as the $\pi^+(u\bar{d})$ and $\pi^-(d\bar{u})$ which were predicted by the Japanese theorist Hideki Yukawa in 1935 as ephemeral entities within atomic nuclei that provided the strong attractive force that holds nuclei together. Their subsequent discovery in 1947 brilliantly confirmed this theory. Today we know of their deeper structure, and also have a more profound understanding of the forces at work on quarks, and antiquarks, which build up the mesons and baryons and ultimately atomic nuclei (see Chapter 7).

There are two neutral combinations that we can form: $u\bar{u}$ and $d\bar{d}$. These make the electrically neutral pion π°, and seed another meson, the electrically neutral **eta** η. Why it is that a single quark can grip a single antiquark like this, but that three quarks or three antiquarks are attracted to form baryons or antibaryons, will be described in Chapter 7.

Chapter 5
Accelerators: cosmic and man-made

For more than a century beams of particles have been used to reveal the inner structure of atoms. These have progressed from naturally occurring alpha and beta particles, courtesy of natural radioactivity, through cosmic rays to intense beams of electrons, protons, and other particles at modern accelerators. By smashing the primary beams into a target, some of the energy can be converted into new particles, which can themselves be accumulated and made into secondary beams. Thus beams of pions and neutrinos, as well as other particles called kaons and muons, have been made, along with antiparticles such as positrons and antiprotons. There are even beams of heavy ions—atoms stripped of their electrons—which enable violent collisions between heavy nuclei to be investigated.

Different particles probe matter in complementary ways. It has been by combining the information from these various approaches that our present rich picture has emerged. For much of the 20th century, beams of particles were directed at static targets, but since the 1970s increasingly counter-rotating beams have been brought into head-on collision. The strategy includes making counter-rotating beams of particles and antiparticles, such as electrons and positrons, or protons and antiprotons, and colliding them head on. Beams of electrons or positrons have been collided with beams of protons at HERA in Hamburg to provide fine-grain images of the internal structure of protons. The most extreme examples of colliding beams are at CERN's Large Hadron Collider, where protons or heavy ions collide at energies above a TeV, 1,000 GeV. Such techniques enable questions to be investigated that would otherwise be impossible, as we shall see later.

There has also been a renewed interest in cosmic rays, where Nature provides particles at energies far beyond anything that we can contemplate achieving on Earth. The problem is that such rays come at random, and are much less intense than beams made at accelerators. It was the desire to replicate the cosmic rays under controlled conditions that in the 1950s inspired development of high-energy physics at accelerators. Today we are recognizing that the Big Bang may have made exotic particles, far more massive than we can ever make on earth, but which might arrive in cosmic rays occasionally. We discovered strange particles (see Chapter 8) in cosmic rays, and later made them to order at accelerator experiments; there is hope that similar fortunes might await us.

Stars and supernovae emit neutrinos; special laboratories have been constructed underground to obstruct the arrival of all but the most penetrating particles, such as neutrinos. Neutrino astronomy is a new area of science that is beginning to flower. One example is the use of the ice pack in Antarctica as a natural detector of cosmic neutrinos.

These are examples of what is known as non-accelerator physics, where natural processes have produced the particles and we detect their effects. Here on Earth we can make intense beams of high-energy particles in laboratories with particle accelerators. In this chapter I shall focus on how accelerators have developed and what is involved in making them.

Electrically charged particles are accelerated by electric forces. Apply enough electric force to an electron, say, and it will go faster and faster in a straight line. In the 1970s the 3-km-long linear accelerator at Stanford in California could accelerate electrons to energies of 50 GeV. It was the world's largest linear accelerator, but it is no longer used for particle physics.

Under the influence of a magnetic field, the path of a charged particle will curve. By using electric fields to speed them, and magnetic fields to bend their trajectory, we can steer particles round circles over and over again. This is the basic idea behind huge rings, such as the 27-km-long Large Hadron Collider at CERN in Geneva.

From cyclotrons to synchrotrons

Exploration of the atom had begun with beams of alpha and beta particles from radioactive bodies. But the individual particles had small energies and restricted ability to get inside the nuclear environment. Beams of high-energy particles changed all that.

The original idea had been to accelerate particles to high energy through a series of small pushes from relatively low accelerating voltages. Particles travel through a series of separate metal cylinders in an evacuated tube. Within the cylinders there is no electric field, and the particles simply coast along. But across the gaps between the cylinders electric fields are set up by means of alternating voltages, which switch between positive and negative values. The frequency of the alternating voltage is matched with the length of the cylinders, so that the particles always feel a kick, not a brake, as they emerge into a gap. In this way, the particles are accelerated every time they cross between one cylinder and the next. This is the basis of the operation of modern linear accelerators. Usually such 'linacs' are short, low-energy machines, but they can be high energy and lengthy, as at the Stanford Linear Accelerator in California. They are most commonly used in the preliminary stages of acceleration at today's big rings, as at the Large Hadron Collider (page 61).

The idea of creating a ring-shaped accelerator originated with Ernest Lawrence, who used a magnetic field to bend the particles into a circular orbit. Two hollow semi-circular metal cavities, or 'Ds', were placed facing one another to form a circle, with a small gap between the two flat faces of the Ds. The whole construction was only about 20 cm across, and Lawrence placed it between the circular north and south poles of an electromagnet, to swing the particles round the curve, while an electric field in the gap provided them with speed. After being accelerated by the electric field in the gap, they curved round in a circular path until they met the gap half an orbit later. By this device they could pass across the same accelerating gap many times, rather than travel through a succession of gaps. They spiral outwards as their speed increases, but the time intervals between successive crossings of the gaps remain constant.

To accelerate the particles continuously, the electric field in the gap must switch back and forth at the same frequency with which the particles complete the circuit. Then particles issuing from a source at the centre of the whirling device would spiral out to the edge and emerge with a greatly increased energy.

This device was known as a 'cyclotron', and worked on the principle that the particles always take the same time to complete a circuit. This is, however, only approximately true in practice. As the energy of the particles increases, the effects of special relativity play an ever more important role. In particular, there is an increasing resistance to acceleration, where more force is required to obtain the same acceleration as the speed approaches that of light. The accelerated particles take longer to complete a circuit, eventually arriving too late at the gap to catch the alternating voltage during the accelerating part of its cycle.

The solution was to adjust the frequency of the applied voltage so that it remains in step with the particles as they take longer to circulate. However, there is a catch: a machine operating at variable frequency can no longer accelerate a continuous stream of particles, as the cyclotron had done. Changing the frequency to keep in time with higher-energy particles would mean that any particles still at lower energies would become out of step. Instead, the 'synchrocyclotron' takes particles from the source a bunch at a time, and accelerates these bunches out to the edge of the magnet.

The synchrocyclotron was able to accelerate protons to sufficient energies that collisions with nuclei produce pions, the lightest particles that, we now know, are made from a single quark and an antiquark. However, the machine was nearly 5 m in diameter, and to go to higher energies, such as those needed to produce the more massive strange particles, was impractical.

The solution was to increase the strength of the magnetic field continuously as the circling particles gain energy, thereby keeping them on the same orbit instead of spiralling outwards. Moreover, the enormous single magnet of the cyclotron can be replaced by a doughnut-like ring of smaller magnets, which is the shape familiar to modern accelerator rings. The particles travel through a circular evacuated pipe held in the embrace of the magnets; they are accelerated during each circuit by an alternating voltage of varying frequency, which is applied at one or more places around the ring; and they are held on their circular course through the pipe by the steadily increasing strength of the magnetic field. Such

11. First successful cyclotron, built in 1930. Lawrence's original cyclotron was only 13 cm in diameter. The magnetic field that steers the particles on a circular path is supplied by two electromagnets. These generate a vertical north–south field through the path of the particles, which are contained in a horizontal plane. They are accelerated by an electric field, which is provided across a gap between the two hollow D-shaped metal vacuum chambers. A radioactive source at the centre provides the particles. The particles curl round in the cyclotron's magnetic field, but as they increase in energy they curl less and so spiral outwards until they emerge from the machine.

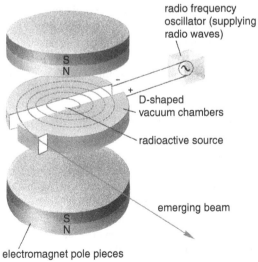

radio frequency oscillator (supplying radio waves)

D-shaped vacuum chambers

radioactive source

emerging beam

electromagnet pole pieces

S N

S N

53

a machine is called a synchrotron. The first major synchrotrons were at Brookhaven in the USA and CERN in Geneva, with energies up to 30 GeV by 1960.

In the 1960s the idea of quarks emerged, and with this came the challenge to reach energies above 100 GeV in the vain hope of knocking quarks out of protons. Improvements in technology led to more powerful magnets, and by placing them in a ring with a diameter of over a kilometre, by the middle of the 1970s Fermilab near Chicago in the USA and CERN had achieved proton energies of some 500 GeV. By 1982 Fermilab had achieved 1,000 GeV, or '1 TeV', and became known as the 'Tevatron'.

Today, superconducting magnets enable even more powerful magnetic fields to be achieved. At Fermilab, alongside the Tevatron, is a smaller ring known as the Main Injector. One of the Main Injector's tasks is to direct protons at 120 GeV onto targets to create secondary beams of particles for experiments. The extracted protons strike special targets of carbon or beryllium to produce showers of pions and kaons. The pions are allowed to decay to produce a neutrino beam, while the kaons can be separated out to form a kaon beam for experiments. Different particles with different properties can probe different features of the target and help to build a richer picture of its make-up.

The Main Injector also directs 120 GeV protons onto a special nickel target at energies sufficient to produce further protons and antiprotons at a rate of up to 200 billion antiprotons in an hour. Antiprotons, the antimatter versions of protons, have negative rather than positive electric charge, and this means that they can travel round the Tevatron's ring of superconducting magnets at the same time and at the same velocity as the protons, but in the opposite direction. Once the particles are at 1,000 GeV, or 1 TeV, the two beams are allowed to collide head on—and the Tevatron has reached its final goal: collisions of protons and antiprotons at

12. The Cosmotron at the Brookhaven National Laboratory, New York. This was the first proton synchrotron to come into operation, in 1952, accelerating protons to an energy of 3 GeV. The magnet ring was divided into four sections (the nearest is clearly visible here), each consisting of 72 steel blocks, about 2.5 m × 2.5 m with an aperture of 15 cm × 35 cm for the beam to pass through. The machine ceased operation in 1966.

energies that recreate the conditions of the universe when it was less than a trillionth of a second old.

The Tevatron was the world's most powerful accelerator until the Large Hadron Collider (LHC) began operation at CERN in 2008. In the LHC, a 27-km ring of magnets steers two counter-rotating beams of protons, or of atomic nuclei, to meet head on at energies up to 7 TeV. This is currently the pinnacle of colliding beam technology.

Linear accelerators

In the 1960s, the Stanford Linear Accelerator became the longest linac in the world. It accelerated electrons to 50 GeV energy in just 3 km, whereas in the 1990s at LEP, a circular accelerator, they reached 100 GeV but required 27 km circumference for the ring. Why this difference, and what decides whether to make a linear or circular accelerator?

Electron synchrotrons work perfectly well apart from one fundamental problem: high-energy electrons radiate away energy when they travel on a circular path. The radiation—known as synchrotron radiation—is greater the tighter the radius of the orbit and the higher the energy of the particle. Protons also emit synchrotron radiation, but because they are 2,000 times as massive as electrons, they can reach much higher energies before the amount of energy lost becomes significant. But even at only a few GeV, electrons circulating in a synchrotron radiate a great deal of energy, which must be paid for by pumping in more energy through the radio waves in the accelerating cavities. It was for these reasons that until recently high-energy electron accelerators have been linear. Indeed, electrons have only been used in circular machines for the special advantages that can arise. Specifically, the head-on collisions make more efficient use of the energy than when a static target is hit. The second major advantage is the ability to probe in ways that would otherwise prove impossible, as

13. A view inside the 27-km (17-mile) circular tunnel of CERN's Large Hadron Collider (LHC). Two beams of protons travel in opposite directions through hundreds of specially designed magnets at energies up to 6.5 TeV.

for instance at LEP—the Large Electron Positron collider, at CERN—where electrons annihilated with positrons, and the counter-rotating beams were the only effective way to achieve the required high intensity.

LEP was a circular machine in a tunnel 27 km long. This is testament to the problems with lightweight electrons and positrons travelling round circles, that such a distance is needed to enable them to reach 100 GeV without wasting too much energy in radiation. To boost electrons to energies of several hundred GeV in circular orbits would require distances of hundreds of kilometres, which are impractical. This is why longer-term plans for very high energy accelerators of electrons and positrons focus on linear colliders.

To have a decent chance of a collision in a linear accelerator, where the beams meet once only, requires high-intensity beams

14. The 3-km- (2-mile-) long linear accelerator at the Stanford Linear Accelerator Center (SLAC). The electrons start off from an accelerator 'gun' where they are released from a heated filament at the end of the machine at the bottom left of the picture. The electrons in effect surf along radio waves set up in a chain of 100,000 cylindrical copper 'cavities', about 12 cm in diameter. The machine is aligned to 0.5 mm along its complete length and situated in a tunnel 8 m below ground. The surface buildings that mark out the line of linac contain the klystrons which provide the radio waves.

that are less than a micron (10^{-6} m) across. In actuality it is more miss than hit. As the like charges within each beam repel one another, making and controlling such tightly focused beams is a technological challenge. If the beams were tuned to each have energy of 62.5 GeV, which is half the mass (rest energy) of a Higgs boson, it could be possible to produce a Higgs boson at rest. This could be a key strategy for a dedicated 'Higgs factory', producing large numbers of Higgs bosons for precision study.

Colliders

In a linear accelerator aimed at a static target the debris of the collision is propelled forwards, just as a stationary car is shunted forwards when another car crashes into its rear. When a beam hits a stationary target, the hard-won energy of the beam particles is being transferred largely into energy of motion—into moving particles in the target—and is effectively wasted. This problem is overcome if we can bring particles to collide head on, so that their energy can be spent on the interaction between them. In such a collision the debris flies off in all directions, and the energy is redistributed with it—none is 'wasted' in setting stationary lumps in motion.

These arguments were clear to accelerator builders as long ago as the 1940s, but it took 20 years for particle colliders to take shape, and another 15 years for them to become the dominant form of particle accelerator.

Protons are bunches of quarks, and antiprotons are likewise made of antiquarks. With a mass of nearly 2,000 times that of an electron, protons and antiprotons suffer less synchrotron radiation and also pack a bigger punch. Hence they are the prime choice when the aim is to reach out to previously unexplored higher energies. Such was the case in 1983 when head-on collisions between protons and antiprotons at CERN led to the discovery of the $W^{\pm}$ and Z^{o} carriers of the weak forces (see Chapter 7).

The proton's energy is shared among its quarks, and it is chance whether the energy of a single quark that meets an antiquark matches that required to form a Z^o or $W^\pm$. Nonetheless they showed up as one in a million special cases in the collection of images of the collisions. The challenge was then to make a Z^o regularly without the vast unwanted and confusing background. This was done by tuning a beam of electrons and positrons to the required energy. This led to the Large Electron Positron collider. The technical challenges of making experiments with such machines can be illustrated by reference to LEP.

When LEP began running, in the 1990s, needle-like bunches of electrons and positrons would pass through each other at the heart of the detectors every 22 microseconds (22 millionths of a second). Even though there were some million million particles in each bunch, the particles were thinly dispersed, so interactions between them were rare. An interesting collision, or 'event', only occurred about once every 40 times or so the bunches crossed. The challenge was to identify and collect the interesting events, and to not miss them while recording something more mundane. An electronic 'trigger' responded to the first signals from a collision to 'decide' within 10 microseconds whether something interesting had occurred. If it had, the process of reading out and combining the information from all the pieces of the detector would begin, and a display on a computer screen would recreate the pattern of particle tracks and show where energy had been deposited in the detector. LEP was closed in 2000 to make way for the Large Hadron Collider (LHC).

From 1992 to 2007, in Hamburg, there was a unique asymmetric collider where a beam of protons collided with a beam of electrons or of positrons. The resulting collisions enabled the proton substructure, and that of its quarks, to be probed at distances down to 10^{-19} m. The resulting information about the quark and gluon structure within a proton proved key to planning experiments with protons at the LHC.

The Large Hadron Collider

The LHC at CERN, commissioned in 2008, is the culmination of decades of accelerator development. To collide proton beams at energies of 7 TeV requires a range of accelerators. The story starts with a bottle of hydrogen gas, about the size of a fire extinguisher. The bottle contains enough protons to run the LHC for decades, though for safety reasons the bottle is replenished twice a year.

First, hydrogen atoms are fed into a linear accelerator. Intense electric fields strip the hydrogen atoms' proton of its surrounding electron. The protons are then accelerated to a speed of about 1/3 that of light. When a carefully monitored number has been collected, the packet is automatically injected into a booster accelerator and their speed increased to about 90 per cent that of light. Straight acceleration is now impractical, so magnets push the beams around a circle of about 160 m circumference while electric fields in regular pulses thrust them, analogous to pushing a child on a swing.

Next, they enter a proton synchrotron. At 630 m in circumference, this was CERN's state of the art accelerator in the 1960s. The protons exit this synchrotron with an energy of about 25 GeV and enter a larger synchrotron which raises their energy to about 450GeV. This, the machine that discovered the W and Z bosons in 1983, is today the launch pad into the LHC itself.

There are two vacuum tubes in the LHC, one for each proton beam. One beam circulates clockwise, the other anticlockwise, around the 27-km ring, 11,000 times each second. At four points around the ring, the beams cross one another.

Each beam consists of packets of protons about 7 m apart in pencil-thin bunches about 60 cm long. In other words, at any moment, most of the tube is empty space. Electronics are tuned so these counter-rotating bunches reach the cross-over points with a precision of better than one tenth of a billionth of a second.

Their energy is raised to as much as 6 TeV. Collisions occasionally produce Higgs bosons. Each collision point is surrounded with cylinders of sophisticated electronics to detect the results of the collisions and tease out evidence of Higgs bosons and other transient particles.

Factories

The conundrum of how matter and antimatter differ has moved into focus in recent years. This has led to an intense interest in the properties of strange particles and antiparticles—the kaons—where a subtle asymmetry was discovered nearly 50 years ago, and their bottom analogues (see p. 101), where a large asymmetry was predicted. This led to the concept of particle 'factories', capable of producing as many kaons or B mesons as possible.

The idea is to make electrons and positrons collide at specific energies, 'tuned' to produce kaons or B mesons, respectively, in preference to other kinds of particles.

A 'B factory' makes electron-positron collisions at a total energy of around 10 GeV, optimized to produce B mesons and their antiparticles ($\bar{B}$, pronounced B-bar) together. So compelling is the challenge that two machines were built in the late 1990s—PEP2 at Stanford in California, which ended in 2008, and KEKB at the KEK laboratory in Japan, which has been superseded by SuperKEKB.

The B factories differ from previous electron-positron colliders in an intriguing way. In a standard electron-positron collider, the beams travel in opposite directions but with the same speed, so that when particles meet their motion exactly cancels out. The resulting 'explosion' when the electrons and positrons mutually annihilate is at rest, and newly created particles of matter and antimatter emerge rather uniformly in all directions. In the B factories, the colliding beams move with different speeds—at SuperKEKB electrons have 7 GeV energy while positrons are at 4 GeV—so the resulting explosion is itself moving.

As a result of this asymmetric collision, the matter and antimatter that emerge tend to be ejected in the direction of the faster initial beam, and at higher speeds than from an annihilation at rest. This makes it easier to observe not only the particles created, but also the progeny they produce when they die—thanks to an effect of special relativity (time dilation) which means that particles survive longer and travel further (about 1 mm) when moving at high speed. These are essential tricks because a B meson, at rest, lives only for a picosecond, a millionth of a millionth of a second, and this is on the margins of measurability.

Intense sources of neutrinos can enable study of these enigmatic particles. Their masses are too small to measure, but indirect measures of their differences in mass can be obtained. There is even the possibility that neutrinos and antineutrinos might change into one another, a form of matter into antimatter that could have important implications for our understanding of this profound asymmetry. The dedicated study of neutrinos is a rich field and is described in Chapters 9 and 10.

Finally, the discovery in 2012 of the Higgs boson, with a mass of 125 GeV, among the debris from collisions between protons at CERN's Large Hadron Collider, has created interest in producing large numbers of these bosons under more controlled conditions. The Chinese have made a proposal to build a 100-km-long tunnel housing an electron-positron collider capable of producing millions of Higgs bosons. Plans for a 'Higgs factory' are still being debated and are unlikely to develop for some decades, however.

Chapter 6

Detectors: cameras and time machines

Early methods

Ways of detecting subatomic particles are more familiar than many people realize. The crackle of a Geiger counter, and the light emitted when electrically charged particles, such as electrons, hit specially prepared materials forming the picture on a classic television screen, are but two.

Rutherford discovered the atomic nucleus by its effect on beams of alpha particles; they had scattered through large angles. He had used scintillating materials to reveal them as they scattered from the atomic nucleus. Rutherford and his colleagues had to use their own eyes to see and count the flashes; by the 1950s electronic components had automated the process of counting the flashes from modern scintillators.

When a charged particle travels through a gas, it leaves behind a trail of ionized atoms. A whole range of particle detectors, from the cloud chamber to the wire spark chamber, depended on sensing this trail of ionization in some way.

By such means, nearly a century ago, Rutherford was able to detect alpha particles that had been emitted by radium or polonium, one at a time.

The key feature was that the detector could greatly amplify the tiny amount of ionization caused by the passage of a single alpha particle. It consisted of a brass tube, which was pumped out to a low pressure, and had a thin wire passing along the centre. The wire and tube had 1,000 volts applied between them, which set up an electric field. Under this arrangement, when a charged particle passes through the rarefied gas, ions are created. They are attracted towards the wire, and as they speed up they ionize more gas, amplifying the initial effect. One ion could spawn thousands, which all end up at the central wire, producing a pulse of electric charge large enough to be detected by a sensitive electrometer connected to the wire.

In the 'Geiger counter' the electric field at the wire is so high that a single electron anywhere in the counter can trigger an avalanche of ionization, such that the tiniest amount of ionization produces a signal.

Although this reveals the presence of radiation, it is far removed from what is needed for detecting particles in modern high-energy experiments. Let's see how detection has developed.

The first detector capable of revealing trails of charged particles was the cloud chamber, which is a glass chamber fitted with a piston and filled with water vapour. When you quickly withdraw the piston, the sudden expansion will cool the gas and a mist forms in the cold, damp atmosphere. When alpha and beta particles from radioactivity pass through, they ionize atoms in the vapour and cloud drops form instantly around their trail. When illuminated, the tracks stand out like the dust motes in a sunbeam.

Historically, mostly in the first half of the 20th century, the cloud chamber was used to detect particles in cosmic rays, its efficiency improved by combining it with the Geiger counter. Put one Geiger counter above and another below the cloud chamber, then if both fire simultaneously it is very likely that a cosmic ray has passed through them and, by implication, through the chamber. Connect the Geiger counters to a relay mechanism so that the electrical impulse from their coincident discharges triggers the expansion of the cloud chamber, and a flash of light allows the tracks to be captured on film.

The first example of an antiparticle, the positron, and also strange particles were discovered in cosmic rays by means of the cloud chamber. However, such techniques were superseded by the use of emulsions.

Emulsions

Photographic plates had figured in the very earliest work on radioactivity; indeed, it was through the darkening of plates that both X-rays and radioactivity were discovered.

In the late 1940s, high-quality photographic emulsions became available. When taken to high altitudes by balloons, they produced the first beautiful images of the interactions of cosmic rays.

These emulsions were especially sensitive to high-energy particles; just as intense light darkens photographic plates, so can the passage of charged particles. We can detect the path of a single particle by the line of dark specks that it forms on the developed emulsion. The particle literally takes its own photograph. A set of emulsion-covered plates is sufficient to collect particle tracks; a cloud chamber, on the other hand, is a complex piece of apparatus, needing moving parts so that the chamber can be continually expanded and recompressed. As a result, emulsions became a useful way of detecting and recording the trails of charged particles.

Bubble chamber

The advent of accelerators produced high-energy particles, which created new challenges for detection. Energetic particles fly through a cloud chamber without interacting with the atoms in the chamber's thin gas. For example, to record the whole life of a strange particle, from production to decay, at energies of a few GeV would have required a cloud chamber 100 metres long! In addition, cloud chambers are slow: the cycle of recompression after an expansion can take up to a minute; by the 1950s, particle accelerators were delivering pulses of protons every two seconds.

What was needed was a detector that would capture the long tracks of high-energy particles and operate quickly. Gases were much too tenuous for the job whereas liquids were better, because their much greater density means they contain far more nuclei with which the high-energy particles can interact. This brings us to the bubble chamber. The basic idea develops from what happens when you keep a liquid under pressure, very close to its boiling point. If you lower the pressure in these circumstances, the liquid begins to boil, but if you lower the pressure very suddenly, the liquid will remain liquid even though it is now above its boiling point. This state is known as 'superheated liquid' and because it is unstable, it can be maintained only so long as no disturbance occurs in the liquid.

Release the pressure and then immediately restore it. Particles entering the liquid during the critical moments of low pressure create a disturbance and trigger the boiling process as they ionize the atoms of the liquid along their paths. For a fraction of a second, a trail of bubbles forms where a particle has passed, which can be photographed. The immediate restoration of pressure would mean that the liquid was once again just below boiling point, and the whole process could be repeated quite rapidly.

The operation of a bubble chamber was always intimately tied to the operating cycle of the accelerator that fed it. The particles entered the chamber when the piston was fully withdrawn, the pressure at its minimum, and the liquid superheated. Then, about 1 millisecond later, an arc light flashed, illuminating the trails of bubbles formed by charged particles. The delay between minimum pressure and the flash allowed the bubbles to grow large enough to show up on the photographs. Meanwhile, the piston moved back in towards the chamber, increasing the pressure again, and the film in the cameras was automatically wound on to the next frame. It then took about a second for the chamber to 'recover' and be ready for the next expansion. Thus the bubble chamber shows where the particles have been, enabling their behaviour to be studied at leisure.

In a magnetic field, a charged particle's trajectory will curve, the direction revealing whether the particle was positively or negatively charged, and the radius of the curve revealing its momentum. So we can deduce the charge and momentum; if you know a particle's momentum and velocity, you can calculate its mass and hence its identity.

One method of pinpointing the velocity used two scintillation 'counters', which produced a flash of light each time a charged particle passed through. Each tiny burst of light was converted to a pulse of electricity, which was then amplified to produce a signal. In this way, two or more scintillation counters could reveal the flightpath of a particle as it produced flashes in each counter, and from the time taken to travel between the two counters, the particle's speed could be determined.

However, such techniques did not help solve the identification puzzle in the case of a bubble chamber picture. Often the only way was to assign identities to the different tracks, and then to add up the energy and momentum of all the particles emerging from an interaction. If they did not balance the known values before the interaction, the assumed identities must be wrong, and others must be tested, until finally a consistent picture was found. This was time consuming, but the state of the art around 1960. Identifying particles through such trial-and-error calculations is the kind of repetitive job at which computers excel, and today

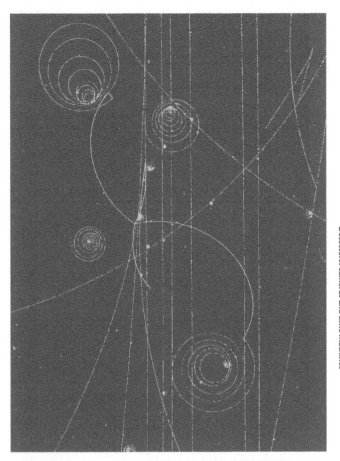

15. Subatomic particles viewed in the bubble chamber at CERN. Cosmic rays provided physicists with the first glimpses of new subatomic particles, which were later studied in detail in experiments at particle accelerators. Positrons, muons, pions, and kaons all figure in this photograph from the 2 m bubble chamber at the CERN laboratory.

bubble chambers have been superseded by electronic detectors that lend themselves better to computer analysis.

From bubble chamber to spark chamber

A bubble chamber could provide a complete picture of an interaction, but it had many limitations. It is sensitive only when its contents are in the superheated state, after the rapid expansion. Particles must enter the chamber in this crucial period of a few milliseconds, before the pressure is reapplied to 'freeze' the bubble growth.

To study large numbers of rare interactions required a more selective technique. In the 1960s, the spark chamber proved the ideal compromise.

The basic spark chamber consists of parallel sheets of metal separated by a few millimetres and immersed in an inert (less reactive) gas such as neon. When a charged particle passes through the chamber it leaves an ionized trail in the gas, just as in a cloud chamber. Once the particle has passed through, you apply a high voltage to alternate plates in the spark chamber. Under the stress of the electric field, sparks form along the ionized trails. The process is like lightning in an electric storm. The trails of sparks can be photographed, or their positions can even be recorded by timing the arrival of the accompanying crackles at electronic microphones. Either way, a picture of particle tracks can be built up for subsequent computer analysis.

The beauty of the spark chamber is that it has a 'memory' and can be triggered. Scintillation counters outside the chamber, which respond quickly, can be used to pinpoint charged particles passing through the chamber. Provided all this happens within a tenth of a microsecond, the ions in the spark chamber's gaps will still be there, and the high-voltage pulse will reveal the tracks.

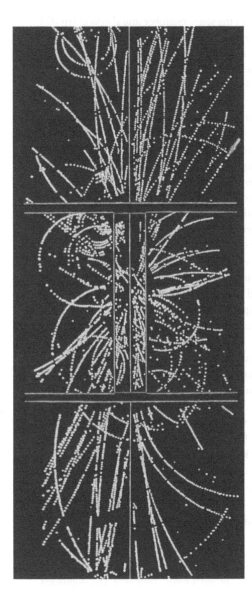

16. An image of one of the first observations of the W particle—the charged carrier of the weak force—captured in the UA1 detector at CERN in 1982. UA1 detected the head-on collisions of protons and antiprotons, which in this view came from the left and right to collide at the centre of the detector. The computer display shows the central part of the apparatus, which revealed the tracks of charged particles throughout the detector. Each dot in the image corresponds to a wire that registered a pulse of ionization. As many as 65 tracks have been produced, only one of which reveals the decay of a W particle created fleetingly in the proton-antiproton collision. The white track (arrowed) is due to a high-energy electron. Adding together the energies of all the other particles reveals that a relatively large amount of energy had disappeared in the direction opposite to the electron, presumably spirited away by an invisible neutrino. Together, the neutrino and electron carry energy equivalent to the mass of the short-lived W particle.

71

Subdivide the plates of the spark chamber into sheets of parallel wires, a millimetre or so apart. The pulse of current associated with each spark is sensed only by the wire or two nearest to the spark, and so by recording which wires sensed the sparks you know to within a millimetre where the particle has passed. Notice how this enables the wire spark chamber to produce information ready for a computer to digest with little further processing.

Wire spark chambers could be operated up to 1,000 times faster than most bubble chambers and fitted in particularly well with the computer techniques for recording data that were developed in the 1960s. Signals from many detectors—scintillation counters, wire chambers—could be fed into a small 'online' computer, which would not only record the data on magnetic tape for further analysis 'offline', but could also feed back information to the physicists while the experiment was in progress. Sets of chambers with wires running in three different directions provided enough information to build up a three-dimensional picture of the particle tracks. And the computer could calculate the energy and momentum of the particles and check their identification.

In the 1960s, spark chambers allowed the rapid collection of data on specific interactions; bubble chambers, on the other hand, gave a far more complete picture of events, including the point of interaction, or 'vertex'. The 'electronic' and 'visual' detectors were complementary, and together they proved a happy hunting ground for the seekers of previously unknown particles. Today working spark chambers are mostly found in science museums and educational establishments to make visible, in particular, the passage of cosmic rays.

Electronic bubble chambers

By the late 20th century, the number of interactions at particle accelerators had become huge compared with those in the days of bubble chambers and even early spark chambers. Developments included the multiwire proportional chamber and the drift chamber, which work much faster and more precisely than wire spark chambers.

A multiwire proportional chamber is superficially rather similar to a spark chamber, being a sandwich of three planes of parallel wires fitted into a gas-filled structure, but differs in that the central plane of wires is held continuously at some 5,000 volts electrical potential relative to the two outer planes. Charged particles then trigger an avalanche of ionization electrons when they pass through the gas. A chamber with wires only 1–2 mm apart produces a signal within a few hundredths of a microsecond after a particle has passed by, and can handle as many as a million particles per second passing each wire—a thousand-fold improvement on the spark chamber.

The downside is that to track particles across a large volume, of a cubic metre say, requires a vast number of wires, each with electronics to amplify the signals. Furthermore, it has limited precision. These problems were overcome with the 'drift chamber', whose basic idea is to measure time—which can be done very precisely with modern electronics—to reveal distance. The chamber again consists of parallel wires strung across a volume of gas, but some of the wires provide electric fields that in effect divide a large volume into smaller units, or 'cells'. Each cell acts like an individual detector, in which the electric field directs the ionization electrons from a charged particle's track towards a central 'sense' wire. The time it takes for electrons to reach this wire gives a good measure of the distance of the track from the sense wire. This technique can locate particle tracks to an accuracy of some 50 microns (millionths of a metre).

Silicon microscopes

Several strange particles live for about 10^{-10} seconds, during which brief span they may be travelling near the speed of light and cover a few millimetres. Over such distances they leave measurable trails. Particles containing charmed or bottom quarks live typically for no more than 10^{-13} s, and may travel only 300 micrometres. To see them one must ensure that the part of the detector closest to the collision point has as high a resolution as possible. Nowadays, almost every experiment has a silicon 'vertex' detector, which can reveal the short kinks where tracks diverge as short-lived particles decay to those with longer lifetimes. The ATLAS and CMS detectors at the Large Hadron Collider, for example, each have over 100 square metres of silicon detectors.

When a charged particle passes through the silicon it ionizes the atoms, liberating electrons, which can then conduct electricity. The most common technique with silicon is to divide its surface during fabrication into fine parallel strips spaced some 20 microns (millionths of a metre) apart, yielding a precision on measuring particle tracks of better than 10 microns.

Silicon strip detectors have come into their own at colliders, providing high-resolution 'microscopes' to see back into the beam pipe, where the decay vertices of particles can occur close to the collision point. They have proved particularly important in identifying B particles, which contain the heavy bottom quark. The bottom quarks prefer to decay to charm quarks, which in turn like to decay to strange quarks. Particles containing either of these quarks decay within 10^{-12} s, and travel only a few millimetres, even when created at the highest energy machines. Yet the silicon 'microscopes' constructed at the heart of detectors can often pinpoint the sequence of decays, from bottom to charm to strange particles. At the Tevatron at Fermilab, the ability to 'see' bottom particles in this way was critical in the discovery of the long-sought top quark, which likes to decay to a bottom quark. The LHCb experiment at the LHC relies on this technology to isolate the decay vertices of bottom particles.

Detecting neutrinos

Any individual neutrino may be very unlikely to interact with matter in a detector, but with enough neutrinos, and large detectors, a few may be caught. The basic idea to detect those rare ones is to exploit their tendency to turn into an electrically charged lepton, such as an electron, when they make a hit, and the electron, being charged, is easy to detect. This is how we have learned a lot about the neutrinos that stream down on us every second from the Sun.

When light passes through material, such as water, it travels slower than when in free space. So although nothing can travel faster than light in a vacuum, it is possible to travel faster than light does through a material. When a particle moves through a substance faster than light does, it can create a kind of shock wave of visible light, known as Cerenkov radiation. The Cerenkov radiation emerges at an angle to the particle's path, and the greater the particle's velocity, the larger this angle becomes. The SuperKamiokande experiment detects neutrinos when they interact in water to make either an electron or a muon, depending on the neutrino's type. These particles, unlike the neutrino, are electrically charged and, moving faster than light through the water, can emit Cerenkov radiation. By carefully analysing the patterns of light, one can distinguish between muons and electrons created in the detector, and hence between muon- and electron-neutrinos.

The Sudbury Neutrino Observatory (SNO) was active until 2006. It was 2,070 metres below ground in a nickel mine in Sudbury, Ontario. Its heart was an acrylic vessel filled with 1,000 tonnes of 'heavy water', called deuterium, in which a neutron joins the single proton of ordinary hydrogen. In SNO, electron-neutrinos interact with the neutrons in the deuterium to create protons and electrons, and the fast-moving electrons emit cones of Cerenkov radiation as they travel through the heavy water. The Cerenkov light forms patterns of rings on the inner surface of the water tank, where it is picked up by thousands of phototubes arrayed around the walls.

However, the key feature of SNO was that it could also detect all three types of neutrinos (see Chapter 8) through a reaction unique to deuterium. A neutrino of any kind can split the deuterium nucleus, freeing the neutron, which can be captured by another nucleus. The capture is detected when the newly bloated nucleus gets rid of its excess energy by emitting gamma rays, which in turn make electrons and positrons that create characteristic patterns of Cerenkov light in the surrounding water.

By such experiments it proved possible to count the neutrinos from the Sun. They confirmed that the Sun is indeed a nuclear fusion engine. That this is how stars, such as the Sun, burn had long been suspected, but it was finally proved in 2002.

Detectors at colliders

Electronic detectors have produced their most spectacular results in an environment that is inaccessible to bubble chambers: at colliding-beam machines where particles meet head on within the beam pipe.

These individual pieces are today combined in cylindrical detectors that surround the interaction point at a collider accelerator. The collision happens on the central axis of the detector. As the debris streams out, it encounters a series of different pieces of detector, each with its own speciality in recognizing particles.

At the Large Hadron Collider bunches of particles pass through each other 40 million times a second, and each time they cross there are over 50 collisions, making more than a billion collisions per second in all. The ensuing data collection rate demanded of the detectors is equivalent to the information processing for 20 simultaneous telephone conversations by every man, woman, and child on Earth.

Huge detectors are housed at the collision points. CMS (Compact Muon Solenoid) and ATLAS (A Toroidal LHC ApparatuS) are exploring the new energy region looking for all kinds of new effects—both expected and unexpected. The ATLAS detector is 5 storeys high (20 m) and yet able to measure particle tracks to a precision of 0.01 mm.

CMS and ATLAS each follow the time-honoured structure for modern particle detectors. First comes the logically named 'inner tracker', which records the positions of electrically charged particles to an accuracy of about one-hundredth of a millimetre, enabling computers to reconstruct their tracks as they curve in the intense magnetic fields. The next layer is a two-part calorimeter, designed to capture all the energy of many types of particle. The inner part is the electromagnetic calorimeter, which traps and records the energies of electrons, positrons, and photons.

High-quality lead glass, like the crystal of cut-glass tableware, is often used as a detector because the lead in the glass makes electrons and positrons radiate photons and also causes photons to convert into electron-positron pairs. The net effect is a miniature avalanche of electrons, positrons, and photons, which proceeds until all the energy of the original particle has been dissipated. The electrons and positrons travel faster in the glass than light does, and emit Cerenkov light, which is picked up by light-sensitive phototubes. The amount of light collected bears testimony to the energy of the original particle that entered the block.

Thousands of tonnes of iron are interleaved with gas-filled tubes to pick up protons, pions, and other hadrons—particles built from quarks. This is the 'hadron calorimeter', so called because it measures the energy of hadrons, just as calorimeters in other branches of science measure heat energy. The iron in the calorimeter has a dual purpose: as well as slowing down and trapping the hadrons, it forms part of the electromagnet used to bend the paths of charged particles, revealing their charge and helping to identify them.

An outermost layer consists of special muon chambers, which track muons, the only electrically charged particles that can penetrate this far. The set of detector components form a hermetic system designed to trap as many particles as possible as they emerge from the collisions at the centre. In principle, only the elusive neutrinos could escape completely, leaving no trace at all in any of the detector components. Yet even the neutrinos left a 'calling card', for they escaped with energy and momentum, both of which must be conserved in any interaction.

This entire detector is designed to record the debris from collisions that occur a billion times each second. This is a far cry from the early days of cloud chambers, which could record only once a minute, or even bubble chambers at once a second. Among the debris produced in these collisions, at energies exceeding anything ever measured at an existing particle accelerator, the jewel will be some unexpected phenomenon. The major discovery announced in July 2012 has been the Higgs boson (Chapters 7 and 10). This particle, with a mass of 125 GeV, is produced on average only once in every 20 million million collisions. This means that with up to a billion collisions each second, a Higgs boson has appeared about once a day in each experiment at the LHC, though with improvements in the intensity of the LHC beams this rate will be higher in future. It has been suggested that finding a needle in a haystack is easier than

17. A LEP detector with four scientists setting the scale.

18. Eight toroid magnets of the ATLAS detector at the Large Hadron Collider.

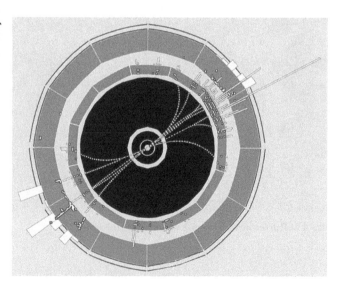

19. Trails of particles and antiparticles as revealed on the computer screen. Compare the computer view with the end view of the detector in Figure 17.

sighting the one Higgs in every hundred thousand billion other events. A challenge for computation has been to recognize the Higgs and record only selected data onto magnetic tape.

This all illustrates how our ability to learn about the origins and nature of matter have depended upon advances on two fronts: the construction of ever more powerful accelerators, and the development of sophisticated means of recording the collisions.

Chapter 7
The forces of Nature

Four fundamental forces rule the universe: **gravity**, the **electromagnetic** force and then two that act in and around the atomic nucleus, known as **strong** and **weak**. The latter pair act over distances smaller than atoms and so are less familiar to our macroscopic senses than are the effects of gravity and magnets. However, they are critical to our existence, keeping the Sun burning and providing the essential warmth for life.

Gravity is the most familiar to us. Between individual atoms or their constituent particles, the effects of gravity are nugatory. The strength of gravity between individual particles is exceedingly small, so small that in particle physics experiments we safely ignore it. It is because gravity attracts everything to everything else that its effects add up until they are powerful, acting over cosmic distances.

Electric forces operate on the familiar maxim 'like charges repel; unlike charges attract'. Thus, negatively charged electrons are held in their paths in atoms by the electrical attraction to the positively charged central nucleus.

Charges in motion give rise to magnetic effects. The north and south poles of a bar magnet are an effect of the electrical motions of atoms acting in concert.

The electromagnetic force is intrinsically much more powerful than gravity; however, the competition between attractions and repulsions neuter its effects over large distances, leaving gravity as the dominant effect at large. However, the effects of swirling electric charges in the molten core of the Earth cause magnetic fields to leak into space. A compass needle will point to the North Pole, which may be thousands of miles distant, due to such effects.

It is the electromagnetic force that holds atoms and molecules together, making bulk matter. You and I and everything are held together by the electromagnetic force. When the apple fell from a tree in front of Isaac Newton, it was gravity that guided it; but it was the electromagnetic force—responsible for making the solid ground—that stopped it continuing down to the centre of the Earth. An apple may fall for many seconds from a great height, accelerated by the force of gravity. But when it hits the floor, it is stopped and turned to pulp in an instant: that is the electromagnetic force at work.

Here is an idea of the relative strengths of the two forces. In a hydrogen atom are a negatively charged electron and a positively charged proton. They mutually attract by their gravity; they also feel the attraction of opposite electrical charges. The latter is 10^{40} times stronger than their mutual gravity. To give an idea of how huge this is, consider the radius of the visible universe: it has been expanding at a fraction of the speed of light, about 10^{16} metres per year, for some 10^{10} years since the Big Bang, so the whole universe is at most 10^{25} metres in extent. The diameter of a single proton is about 10^{-15} metres. So 10^{40} is even bigger than the size of the universe compared to the size of a single proton. Clearly we can safely ignore gravity for individual particles at present energies.

The attraction of opposites holds the electrons in atomic paths around the positively charged nucleus, but the repulsion of like charges creates a paradox for the existence of the nucleus itself. The nucleus is compact, its positive electrical charge due to the many positively charged protons within it. How can these protons, suffering such intense electrical repulsion, manage to survive?

The fact that they do gives an immediate clue to the existence of a 'strong' attractive force, felt by protons and neutrons, which is powerful enough to hold them in place and resist the electrical disruption. This strong force is one of a pair that act in and around the atomic nucleus. Known as the **strong** and **weak**, their names referring to their respective strengths relative to that of the electromagnetic force on the nuclear scale, they are short-range forces, not immediately familiar to our gross senses, but essential for our existence.

The stability of the nuclei of atomic elements can be a delicate balance between the competing strong attraction and electrical repulsion. You cannot put too many protons together or the electrical disruption will make the nucleus unstable. This can be the source of certain radioactive decays, where a nucleus will split into smaller fragments. Neutrons and protons feel the strong force equally; only the protons feel the electrical repulsion. This is why the nuclei of all elements other than hydrogen contain not just protons, but have neutrons to add to the strong attractive stability of the whole. For example, uranium 235 is so called because it has 92 protons (which define it as uranium due to the 92 electrons that will neutralize the atom) and 143 neutrons, making a total of 235 protons and neutrons in all.

At this point you might wonder why nuclei favour any protons at all, as an excess of neutrons doesn't seem to lead to instability. The answer depends in detail on quantum mechanical effects that are beyond the scope of this book, but a major part is due to the extra mass of a neutron relative to a proton. As we saw earlier, this underlies an intrinsic instability of neutrons, whereby they can decay, turning into protons and ejecting an electron—the so-called 'beta' particle of 'beta radioactivity'.

The force that destroys a neutron is the weak force, so called because it appears weak by comparison to the electromagnetic and strong at room temperatures. The weak force disrupts neutrons and protons, causing the nucleus of one atomic element to transmute into another through beta radioactivity. It plays an important role in helping convert the protons—the seeds of the hydrogen fuel of the Sun—into helium (the process by which energy is released, eventually emerging as sunshine).

The gravitational attractions among the multitudinous protons in the Sun pull them inwards until they are nearly touching. Occasionally two move fast enough to overcome their electrical repulsion momentarily, and they bump into one another. The weak force transmutes a proton into a neutron, the strong force then clumps these neutrons with protons, after which they build up a nucleus of helium. Energy is released and radiated courtesy of the electromagnetic force. It is the presence of these four forces and their different characters and strengths that keep the Sun burning at just the right rate for us to be here.

In ordinary matter, the strong force acts only in the nucleus and fundamentally it is due to the presence of the quarks, the ultimate basic particles from which protons and neutrons are formed. As the electric and magnetic forces are effects arising from electric charges, so is the strong force ultimately due to a new variety of charge, which is carried by quarks but not by leptons. Hence leptons, such as the electron, are blind to the strong force; conversely, particles such as protons and neutrons that are made of quarks do feel the strong force.

The laws governing this are fundamentally similar to those for the electromagnetic force. Quarks carry the new charge in what we can define to be the positive form, and so antiquarks will carry the same amount but with negative charge. The attraction of opposites then brings a quark and an antiquark together: hence the $q\bar{q}$ bound states that we call mesons. But how are baryons, which are made of three quarks, formed?

It turns out that there are three distinct varieties of the strong charge, and to distinguish among them we call them red (R), blue (B), and green (G). As such they have become known as colour charges, though this has nothing to do with colour in its familiar sense—it is just a name. As unlike colours attract, and like repel, so would two quarks each carrying a red colour charge, say, mutually repel. However, a red and a green would attract, as would three different colours, RBG. Bring a fourth quark near such a trio and it will be attracted to two and repelled by the third which carries the same colour charge. The repulsion turns out to balance the net attraction such that the fourth quark is in some sort of limbo; however, should it find two other quarks, carrying each of the two other colour charges, then this trio can also tightly bind together. Thus we begin to see the attractions of trios, as when forming protons and neutrons, is due to the threefold nature of colour charges. As the presence of electric charges within atoms leads to them clustering together to make molecules, so do the colour charges within protons and neutrons lead to the clusters that we know as nuclei.

The underlying similarity in the rules of attraction and repulsion give similar behaviour to the electromagnetic and strong forces at distances much less than the size of an individual proton or neutron; however, the threefold richness that positive or negative colour charges have in comparison with their singleton electric counterparts leads to a different behaviour in these forces at larger distances. The colour-generated forces saturate at distances of around 10^{-15} metres, the typical size of a proton or neutron, and are very powerful, but only so long as the two particles encroach to within this distance—figuratively 'touch' one another—hence the colour-induced forces act only over nuclear dimensions. The electromagnetic force, in contrast, acts over atomic dimensions of some 10^{-10} metres when building stable atoms, and can even

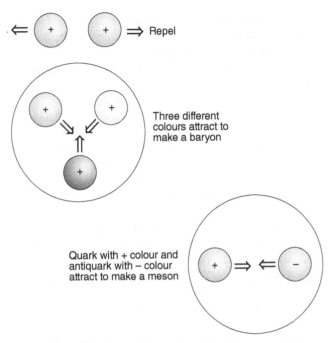

20. Attraction and repulsion rules for colour charges. Like colours repel; unlike colours can attract. Three quarks each carrying a different colour attract to form a baryon. A quark and an antiquark carry opposite colours and can also attract to form a meson.

be felt over macroscopic distances, as in the magnetic fields surrounding the Earth.

This brings us naturally to the question of how forces spread their effects across space.

Force carriers

How do forces, such as the electromagnetic force, manage to spread their effects across space? How does a single proton manage to ensnare an electron that is 10^{-10} metres away, thereby forming an atom of hydrogen? Quantum theory implies that it is by the action of intermediate agents—the exchange of particles; in the case of the electromagnetic force these are photons, quantum bundles of electromagnetic radiation, such as light.

Electric charges can emit or absorb electromagnetic radiation, and its agents, photons; analogously the colour charges also can emit and absorb a radiation, whose agents are known as **gluons**. It is these gluons that 'glue' the quarks to one another to make protons, neutrons, and atomic nuclei. The weak force analogously involves force carriers known as **W** or **Z** bosons.

The photon, W and Z bosons, and the gluons are known as 'gauge bosons'—particle physics' jargon for carriers of fundamental forces which are described by a common mathematical structure known as 'gauge theory'. Gauge bosons carry one unit of spin, making them 'vector' particles. (The Higgs boson is not a 'gauge boson'; it has no spin, uniquely so among the known fundamental particles.)

The W boson differs from the photon in two important ways: it has electric charge and a large mass. Its electric charge causes its emission to leak charge away from the source—thus a neutral neutron turns into a positively charged proton when a W^- is emitted; this is the source of neutron beta decay, the W^- turning into an electron and anti-neutrino. The W mass is some 80 times greater than that of a proton or neutron. If you were in a car weighing one tonne and suddenly 80 tonnes were ejected, you would complain that something was wrong! But in the quantum world this kind of thing can happen. However, this violation of energy balance is ephemeral, limited in time such that the product of the imbalance, Delta-E (ΔE), and the time it can last, Delta-t (Δt), cannot exceed Planck's quantum h, or numerically $\Delta E \times \Delta t < 6 \times 10^{-25}$ GeV-seconds. This restriction is one form of the 'Heisenberg Uncertainty Principle'.

This means that for one second you could overdraw the energy account, or 'borrow', the trifling amount of 10^{-25} GeV. 'Borrowing' 80 GeV (the minimum energy to make a single W) can occur for some 10^{-24} s, during which time not even light could travel more than about one tenth of the distance across a proton. Hence the distance over which the W can transmit the force is considerably less than the size of a single proton. So the short-range nature of the weak force is due to the excessively large mass of its carrier particle. Now, this is not a statement that the force exists up to a certain point and then turns off suddenly; instead it dies away, and its strength falls away radically by distances of the order of a proton size. It is at such distances where beta decay is manifested, and it is thus that the force became known as 'weak'.

In 1865 James Clerk Maxwell had successfully unified the disparate phenomena of electricity and magnetism into what we today call electromagnetism. A century later, Glashow, Salam, and Weinberg united the electromagnetic force and the weak force into what has become known as electroweak theory. This explained the apparent weakness of the 'weak' component of this unified force as being due to the large mass of the W, whereas the photon of the electromagnetic force is massless. Their theory implied that in addition to the electrically charged W^+ and W^-, there was a heavy neutral partner, the Z^0, with a mass around 90 GeV. One implication of their work was that if one could provide enough energy, of the order of 100 GeV or more, whereby the W or Z could be produced directly in the laboratory, one would see that the force has a strength akin to that of the electromagnetic, and is not excessively weak after all. Such experiments have been done and confirmed this phenomenon.

The W and Z were discovered at CERN in 1983–4, where they were fleetingly produced among the debris arising from the high-energy, head-on collisions between protons and antiprotons. Such collisions produce large numbers of pions, and only rarely is

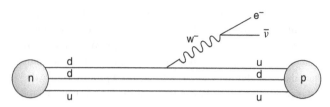

21. Beta decay via W. A neutron converts to a proton by emitting a W, which then turns into an electron and neutrino.

THE FORCES	INTENSITY	CARRIER	EXAMPLE
STRONG	1	GLUONS	atomic nucleus
ELECTROMAGNETIC	$\sim 10^{-2}$	PHOTON	atoms
WEAK	$\sim 10^{-5}$	W^+, W^-, Z^0	neutrinos
GRAVITATION	$\sim 10^{-42}$	GRAVITON	galaxies, planets

22. The relative strengths of the various forces when acting between fundamental particles at low energies typical of room temperature. At energies above 100 GeV, the strengths of the weak and electromagnetic forces become similar. The carriers of the forces are shown: the gluons, photon, and graviton are all massless; the W^+, W^-, and Z^0 are massive. Examples of entities that have special affinity for the various forces are also shown.

a single W or Z produced. This led to a dedicated accelerator, LEP, where counter-rotating beams of electrons and positrons were mutually annihilated, tuned to a total energy of 90 GeV. This energy matches that of a Z at rest, and so LEP was able to produce Z particles cleanly. During a decade of experiments, over 10 million examples of Z were made and studied. These experiments proved that the concept of the merging of electromagnetic and weak forces into a single electroweak force is a correct one. It is the large masses of the W and Z that gave the apparent weakness when they were involved in historical experiments, at energies far below 100 GeV, such as in beta radioactivity.

Particle Physics

Finally we have the strong force, whose origins are the colour charges carried by quarks or antiquarks. In this case the force is transmitted by 'gluons'. As a quark can have any of three colours, labelled R, B, or G, the gluon radiated can itself carry colour charge. For example, a quark with charge R can end up carrying colour B if the gluon carries a charge that is like 'positive R, negative B'. The relativistic quantum theory, known as quantum chromodynamics, or QCD, allows for a total of eight different colours of gluons.

As gluons carry colour charge, they can mutually attract and repel as they travel across space. This is unlike the case for photons when transmitting the electromagnetic force. Photons do not themselves carry (electric) charge, and so do not mutually suffer electromagnetic forces. Photons can voyage across space independently, filling all the volume, the intensity of the resulting force dying out as the square of the distance—the famous 'inverse square law' of electrostatics. Gluons, carrying colour charges, do not fill space in the same way as photons do. Their mutual interactions cause the ensuing force to be concentrated in a line, along the axis connecting the two coloured quarks.

So while photons fill space and travel independently, the gluons cluster. One consequence of this clustering is the possibility that gluons mutually attract to form short-lived composite states known as glueballs. It is this mutual affinity amongst gluons while they are transmitting the force that causes the long-range behaviours of the electromagnetic and colour (strong) force to differ radically. The electromagnetic force dies with the inverse square of the distance; the colour forces do not. The energy required to pull two colour sources, such as quarks, apart grows with distance. At a separation of around 10^{-15} metres this energy tends to become infinite. Thus individual quarks cannot be separated from their siblings; they remain clustered in the trios, such as baryons, or quark and antiquark, as in mesons. It is thus that the effects of colour charges become 'strong' at large distances.

At short distances, as are probed in experiments at high energy, the electroweak and colour forces appear to act as if exhibiting a grand unity. As electrically charged particles can radiate photons, so can coloured quarks radiate gluons. This has been seen in '3-jet events' at LEP, as illustrated in Figure 23. It is at lower energies, such as were the norm until the latter part of the 20th century, that they exhibit their different characters: the massive W and Z causing an apparent weakness; the mutual

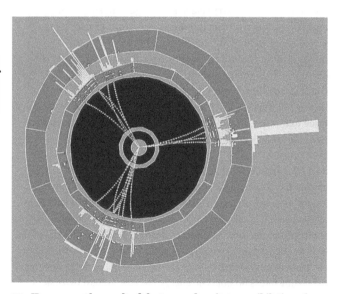

23. Here we see the result of electron and positron annihilation where three jets of particles have emerged. First a quark and an antiquark were produced, and almost immediately one of these radiated a gluon. The quark, antiquark, and gluon are the sources of the three jets of detected particles.

interactions among the gluons causing the colour force by contrast to take on great strength.

That much we know. Extrapolate the effects of the colour forces, and of the weak and electromagnetic to extreme energies, far beyond what we can measure in the lab, and it appears that all three become alike. The behaviour of atomic particles at very high energies, akin to those that were abundant just after the Big Bang, suggests that the colour forces are enfeebled, and similar in strength to the familiar electromagnetic force. A tantalizing hint of unity has emerged. This is known as grand unification of the forces. It suggests that there is an underlying simplicity, unity, to Nature and that we have only glimpsed a cold asymmetric remnant of it so far. Whether this is really true is for future experiments to test.

The mathematical structures underpinning the 'gauge theories' that describe electromagnetic, weak, and strong—colour—forces can be generalized straightforwardly, but there is no evidence that Nature makes use of this. There could be further forces, mediated by very massive gauge bosons and too feeble to discern in present experiments. The properties of the muon—a seemingly more massive version of the electron (see Chapter 8)—give hints that this might be the case, but this remains highly speculative.

Higgs boson

Even if space were to be emptied completely of matter and all known sources of energy, it would still be filled by a ghostlike field that cannot be shut down. Immersed in this essence forever, we have nonetheless been unaware of it until its existence was established in July 2012 with discovery at CERN's Large Hadron Collider of the Higgs boson. As excitation of an electromagnetic field produces photons, so can excitation of this 'Higgs field' produce Higgs bosons.

Though the Higgs field is not a sibling to the electromagnetic, weak, and strong fields, where forces are transmitted by spin-1, gauge bosons, it is nonetheless fundamental to the formation of structure from the chaotic debris of the Big Bang. A small amount of energy can excite millions of photons, but it takes 125 GeV focused into a tiny volume to excite even one Higgs boson. Whereas the electromagnetic and other fields have a sense of direction—they are vector fields—the Higgs field has none: the Higgs boson has no spin; it is a fundamental scalar boson—the only such example yet known.

But for the presence of this field, the fundamental particles would race through space at the speed of light, without any possibility of being caught in atoms or molecules. Their interaction with the field gives them mass. The mass of the W boson enfeebles the 'weak' force such that the conversion of hydrogen to helium in the Sun takes place very slowly. This has enabled the Sun to last 5 billion years, long enough for evolution to have occurred here on Earth. So, although the Higgs boson and the associated field might appear arcane, they are essential to our existence.

Chapter 8
Exotic matter (and antimatter)

Strangeness

We have met the basic particles from which matter on Earth is ultimately made. However, in Nature's scheme, there is more than this. Cosmic rays from outer space are continuously hitting us. These consist of the nuclei of elements produced in stars and catastrophic events elsewhere in the cosmos; they hurl through space and some, channelled by the magnetic fields of the Earth, hit the upper atmosphere and produce showers of secondary particles. In the 1940s and 1950s, cosmic rays provided an active source of discovery of forms of matter that had not hitherto been known on Earth. Some of these had unusual properties and became known as 'strange' particles. Today we know what distinguishes them from the familiar protons, neutrons, and pions: they contain a new variety of quark, which has become known as the **strange quark**.

There are strange baryons and strange mesons. A strange baryon consists of three quarks, at least one of which is a strange quark; the greater the number of strange quarks the baryon contains, the greater is the magnitude of its 'strangeness'. A meson consists of a quark and an antiquark and so, by analogy, a strange meson is one that contains either a strange quark or a strange antiquark. The discovery of strange particles preceded by several years the discovery that baryons and mesons are made of quarks. The properties of the variety of strange particles led theorists to invent the concept of strangeness, which acted in many ways like charge: strangeness is conserved when the strong force acts on particles. Thus one could explain which processes were favoured or disfavoured by computing how much strangeness each of the participating particles carried. Various mesons were determined to carry strangeness of amount +1 or –1. Strange baryons were found by this scheme to carry amounts –1, –2, or –3. Today we understand what determines this. The amount of *negative* strangeness that a particle carries corresponds to the number of strange quarks within it. It might seem more natural to have defined strangeness such that each strange quark carried one unit of positive strangeness, and had we known of quarks before the idea of strangeness, that is probably how it would have been. But we are stuck with this accident of history whereby the number of strange quarks accounts for negative strangeness and the number of strange antiquarks accounts for positive strangeness. (A similar accident of history gave us a negative charge for the electron.)

A strange quark is electrically charged, carrying an amount –1/3, as does the down quark. It is more massive than a down quark, having an mc^2 of ~150 MeV. In all other respects the strange and down quarks appear to be the same. Due to the extra mass of the strange quark relative to an up or down quark, every time one of these in the proton or neutron, say, is replaced by a strange quark the resulting strange baryon is roughly 150 MeV more massive per unit of (negative) strangeness.

The baryons that are like the proton and neutron, and have spin 1/2, are listed in Figure 24a along with their quark content, electric charge, strangeness, and magnitude of mass (or mc^2 in MeV). The rule is not exact, but it is at least qualitatively true (the actual masses, as was the case for the proton and neutron, depend also on the different electrical forces among the constituents and the fact that their sizes, while approximately 10^{-15} m, are not all identical, due to the complicated nature of the forces acting on them). The rule is more precisely verified in the set of strange baryons with spin 3/2 that partner the Δ resonance, as seen in Figure 24b.

There are mesons with strangeness +1, such as the $K^+(u\bar{s})$ or $K^o(d\bar{s})$, and –1, such as $K^-(s\bar{u})$ or $\bar{K}^o(s\bar{d})$, with masses mc^2 ~500 MeV. There are also mesons which contain both strange quark and antiquark, so that there is no net strangeness. This $s\bar{s}$ combination leads to a third electrically neutral meson, known as the eta-prime, η', in addition to the π^o and η that we met in Chapter 4.

a

baryon	quarks	charge	strangeness	mc^2 (MeV)
proton	uud	+1	0	938
neutron	ddu	0	0	940
Lambda Λ	uds	0	–1	1115
Sigma Σ^+	uus	+1	–1	1189
Σ^0	uds	0	–1	1192
Σ^-	uds	–1	–1	1197
XiΞ^0	uss	0	–2	1315
Ξ^-	dss	–1	–2	1321

b

baryon resonance	quarks	strangeness	mc^2 (MeV)
Δ^-	ddd	0	1232
Σ^{*+}	dds	–1	1380
Ξ^{*-}	dss	–2	1530
Ω^-	sss	–3	1670

24. a) Baryons with spin 1/2. b) Baryons with spin 3/2.

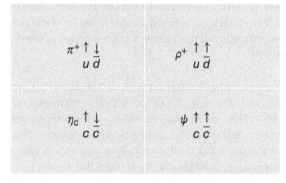

25. Spins of mesons made from quarks. Spins of the u and d quarks add together, forming a positively charged ρ or cancelling out, making a positively charged π. Similar combinations occur for any mixture of u, d, or s flavours with any of their antiquark counterparts. This picture extends to charm, bottom, and top flavours. Among the many resulting combinations we illustrate the 'psi' (ψ), where the spins add to a total of 1, and its partner, the 'eta-charm' η_c, where the spins cancel to zero.

These mesons made of a quark and antiquark have a total spin of zero. There is also a set where the quark and antiquark spins combine to a total of one. The strange members in this case are known respectively as $K^{*+}(u\bar{s})$, $K^{*o}(d\bar{s})$, $K^{*-}(s\bar{u})$, and $\bar{K}^{*o}(s\bar{d})$; the analogues of the π, η, and η' are known as ρ, ω, and ϕ (rho, omega, and phi).

Charm

Not only does the down quark have its heavier cousin, the strange quark, but so does the up quark have a heavier version: the **charm** quark. A charm quark is electrically charged, carrying an amount +2/3, as does the up quark. It is more massive than the up quark, having an mc^2 of ~1500 MeV. In all other respects the charm and up quarks appear to be the same.

In the case of strange quarks, we formed strange baryons and mesons which were a few hundred MeV more massive than their up and down flavoured counterparts. A similar story happens with the charm quark, but due to its greater mass, the analogous charmed mesons and baryons weigh in correspondingly heavier, the lightest being found around ~1900 MeV or nearly 2 GeV. In part as a result of this greater mass, they are not easily produced in cosmic rays, and it was only with the advent of dedicated experiments at high-energy particle accelerators that the existence of charmed particles, and the charm quark, became known in the final quarter of the 20th century.

Charm quarks can link in threes with any combination of up, down, or strange quarks to make baryons with charm, or even with both charm and strangeness. A few examples have even been seen where two charmed quarks have joined with an up, down, or strange quark. We expect that three charmed quarks can join to make a baryon with three units of charm, but clear evidence for its existence is still awaited.

A charmed quark can link with a single antiquark that can be any of (anti)- up, down, or strange. The most celebrated examples, though, are where a charmed quark joins with a charmed antiquark, $c\bar{c}$, leading to yet another electrically neutral partner, adding to the pion and etas, made from $u\bar{u}$, $d\bar{d}$ or $s\bar{s}$ that we already met. The resulting 'eta-c', written η_c, has a mass of just below 3,000 MeV, 3 GeV, and as such is the lightest example of a whole spectroscopy known as 'charmonium'.

It was through charmonium that the charm property was first discovered. The η_c is formed when the c and $\bar{c}$, each having spin 1/2, couple their spins to a total of zero (see Figure 25). They can also couple their spins to give a total value of one; this forms a slightly heavier state at 3.1 GeV known as the psi: ψ. When an electron and a positron meet and annihilate, they do so most

↑ ↑	name	mass (GeV)
$u\,\bar{u}$	$\left.\rule{0pt}{18pt}\right\}\rho^0,\,\omega$	0.8
$d\,\bar{d}$		
$s\,\bar{s}$	ϕ	1.0
$c\,\bar{c}$	ψ	3.1
$b\,\bar{b}$	Υ	9.5
$t\,\bar{t}$	?	370.0 (?)

26. Mesons with spin 1 that can be made easily in e⁺e⁻ annihilation. In addition a photon or Z^0, which are not made from quarks, can be made this way. A tt^* analogue at around 370 Gev probably does not occur, as the top quark and antiquark decay before they can bind to one another.

readily when their spins are correlated to make spin one. In such a reaction both the energy and also the amount of spin are conserved; this has the effect that, if the combined energy of the electron and positron matches the mc^2 of a meson with spin one, made of a quark and its antiquark (hence electrically neutral), then that meson will be produced from the energy left from the annihilation of the electron and positron. So, for example, if an electron and positron collide head on with a combined energy of about 0.8 GeV, which is the mass of the spin-one ρ and ω, either of these mesons can be formed; around 1 GeV the analogous meson made of $s\bar{s}$, namely the φ, appears; and at 3.1 GeV we meet $c\bar{c}$, whence a ψ can be formed. That is how this first example of charmonium was found in 1974, and how the spectrum of particles was gradually uncovered.

Particles with either charm or strangeness are not stable. Their masses are greater than those of baryons or mesons without charm or strangeness, and hence their intrinsic energy, represented by mc^2, is greater. Thus, although strange and charmed particles can be made in high-energy collisions at accelerators, or even in the extreme energies that were prevalent immediately following the Big Bang, they rapidly decay, leaving ultimately up and down quarks within the 'conventional' baryons, which survive in our day-to-day world; mesons ultimately self-destruct due to quark and antiquark annihilation, producing photons or electrons and neutrinos as their stable end products.

Bottom and top

We have seen above how Nature has duplicated its basic quark flavours making a second set, the strange and charm, with the same electric charges but greater mass than their down and up cousins. One may well ask why? This is not the end of the story; Nature has availed itself of a third set of yet more massive quarks, with the same electric charges as those that went before. Thus, we have the bottom quark (b), $mc^2 \sim 4.5$ GeV, electric charge $-1/3$; and there is the top quark (t), $mc^2 \sim 180$ GeV (this is not a misprint!), electric charge $+2/3$. How it is that Nature packs so much mass, comparable to that of an entire atom of gold, into a space of at most 10^{-18} m is one of the great mysteries for the 21st century. In some articles these attributes are called truth and beauty instead of top and bottom; it is the latter that are now rather generally agreed on, and so I shall refer to top and bottom here.

Baryons and mesons containing bottom quarks or antiquarks occur and are in effect heavier analogues of those containing the lighter strange quark of the same charge. The lightest bottom mesons have mass, or mc^2, at around 5 GeV. Similarly, bottom baryons occur. There is little to be gained in writing out all their characteristics; however, if you want to do so, go to the table of strange particles, replace s by b and add about 4.5 GeV mass for every b quark or antiquark, and you will have it. Bottom mesons have proved interesting in that their behaviour may give clues to the puzzle of why the universe is made of matter to the exclusion of antimatter. There is also a spectroscopy of 'bottomonium' states analogous to the charmonium spectroscopy; bottomonium consists of $b\bar{b}$, the lightest example having a mass of around 9.5 GeV.

You might at this point expect that mesons and baryons containing top quarks will occur, with properties analogous to those of the charmed particles (as top and charm have the same charge), and that their main distinguishing feature is that they are nearly 200 GeV more massive than their charmed counterparts. However, there is no evidence for these states. The problem is that the top quark, being so massive, is so unstable that it decays in less than 10^{-25} s. A signal travelling at light speed can only cover about 1/100 fm in this time, too little to grip other quarks or antiquarks to form the bound states that we call mesons and baryons.

Particle Physics

The decay occurs by a process analogous to that familiar in radioactive (beta) decay. As a neutron turns into a proton when a down quark turns into the (lighter) up quark, emitting energy in the form of an electron and a neutrino (technically, an antineutrino),

$$d \to u(e^{+}\bar{\nu})$$

so do the heavier quarks imitate this. The difference between the electric charges of any quarks is either zero or ± 1. In the latter case, a decay can occur from the heavier to the lighter by emitting an electron or a positron respectively (along with a neutrino or antineutrino). So we can have a cascade of decays

$$t \to b(e^{+}\nu); b \to c(\overline{e\nu}); c \to s(e^{+}\nu); s \to u(\overline{e\nu})$$

and at the final step one can have a stable particle left, such as a proton. It is possible, though less likely, that a decay chain might miss a step, e.g. $t \to d(e^{+}\nu)$ or $b \to u(e^{-}\nu^{-})$. It is also quite probable that the charmed quark takes an alternate route

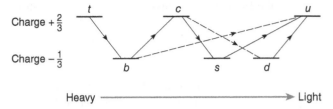

27. **Dominant weak decays of quarks. Each downward arrow emits**
$e^+\nu$; **each upward arrow emits** $e^-\nu^-$. **Two less probable paths are also**
shown with dotted arrows.

$c \to d(e^-\nu^-); d \to u(e^+\nu)$. The d and u quarks have such similar masses, reflected in the similar masses of the neutron and proton, that the process $d \to u(e^-\nu^-)$ is slow, for example the half-life of a free neutron is as long as ten minutes. The other mass differences are larger, and the processes occur faster, in the case of the top so fast that top mesons and baryons do not have time to form.

Who ordered that?

Our world consists of up and down quarks, the electron, and a neutrino. The latter is known as the 'electron-neutrino', symbol ν_e to denote the fact that it is a sibling of the electron. Nature triplicates the quarks, with charm and strange, and also top and bottom as heavier versions of these electrically charged +2/3 and –1/3 particles. It is not just with the quarks that Nature does this; there are three varieties of each of the leptons too.

There is a heavier version of the electron, known as the muon, symbol μ^-. This is negatively charged, like the electron. The muon (and its antiparticle version the μ^+) are apparently in all respects the same as electrons or positrons except that they are 207 times more massive, with an $mc^2 \sim 105$ MeV. In weak decays, the muon is accompanied by a neutrino, but a different neutrino from the ν_e. We call this the muon-neutrino, symbol ν_μ (there is, of course, an antineutrino too: $\bar{\nu}_\mu$).

There is a third set of leptons. This consists of the tau, a negatively charged analogue of the electron but weighing in at some 2 GeV (this is denoted τ^-, its antiparticle version being τ^+), and the associated neutrino (antineutrino) being ν_τ ($\bar{\nu}_\tau$).

Leptons appear to be fundamental. The muon couples to the Higgs boson some 207 times as strongly as an electron does, making it correspondingly more massive. If the muon were just a heavy electron, it would be able to shed 206 parts of that mass as energy in the form of photons, tumbling down to stability: an electron. No sign of this transition has ever been seen, however. There appears to be something intrinsic to the muon and to the electron that must be preserved. This is the property we call flavour; it has superb empirical provenance, but we have no understanding of what underpins it.

In April 2021, the magnetic moment of the muon was measured at Fermilab to an accuracy of ten decimal places. At this level of precision there appears to be a disagreement between experiment and theory, but not yet at a level to claim a discovery. If this discrepancy survives, it could be a hint of virtual massive particles disturbing the vacuum and/or further forces, ultra-weak at present energies, affecting the muon's magnetism.

Neutrinos

In the standard model, neutrinos are assumed to have no mass. This was because historically no one had ever been able to measure a value for any mass that they might have, the amount being so tiny that it might well have been zero. However, there is no fundamental principle that requires neutrinos to be massless. And indeed, we now know that neutrinos do have a mass, exceedingly small compared to even the electron mass, but non-zero nonetheless.

There are three known varieties of neutrino, the electron-neutrino, muon-neutrino, and tau-neutrino, named for their affinity for being produced in concert with the electrically charged particle that shares their name—the electron (e), mu (μ), and tau (τ) leptons. I will refer to these as nu-e, nu-mu, and nu-tau, respectively. They are traditionally denoted by the Greek symbol for nu and the appropriate subscript, thus ν_e, ν_μ, or ν_τ. The fusion reactions in the heart of the Sun emit neutrinos of the nu-e (ν_e) variety.

When they interact with matter and convert neutrons into protons, for example, they turn back into e, mu, and tau. For many years this was thought to be absolutely true, but today we know there is a small chance of a neutrino produced as ν_e ending up as ν_μ or ν_τ. This is due to a phenomenon known as neutrino oscillations and is an indicator that neutrinos have masses, tiny indeed, but not zero.

In quantum mechanics, particles have wavelike character. As the oscillations of the electromagnetic field can take on particle characteristics—the photons—so do particles such as neutrinos have wavelike oscillations as they travel through space. In effect it is a wave of varying probability. What set out as a nu-e will vary in probability as it travels, changing from nu-e to nu-mu or nu-tau as it moves away from the source. However, for this to happen, the neutrinos must have different masses, which implies that not all of them can be massless.

If physical neutrinos are a mixture of two or more underlying quantum states with different masses, then for a given energy the quantum waves of these underlying states will have different frequencies and so swell back and forth into one another. The smaller their mass difference, the more alike their frequencies, the slower their oscillation and the longer the distance needed for any effect to be detectable (see Figure 28).

As neutrino masses are themselves very small, any differences must also be tiny and oscillation lengths large. The first hints

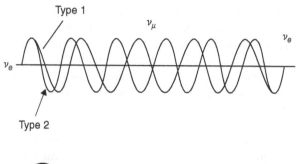

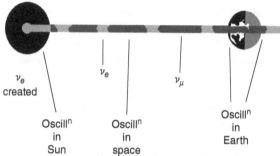

28. **Neutrino oscillations.** (a) The upper figure shows a ν_e composed of two mass states $\nu_{1,2}$, which oscillate at different rates. When they are out of phase, as in this example, it corresponds to only ν_μ being present, the ν_e having vanished. (b) In the lower figure, a ν_e formed in the Sun may oscillate back and forth during its transit through the Sun, through space, and even through the Earth (at night). A comparison of the ν_e intensity by day and by night can reveal the effect of their passage through the Earth.

came in the late 1960s with detection of neutrinos from the Sun. The fusion processes in the Sun produce ν_e. Over several decades, the intensity of nu-e arriving from the Sun was measured. Given our knowledge of the way that the Sun works, it was possible to compute the number of ν_e it produced and hence the intensity of them when they reach the Earth. However, when the measurements were made, the intensity of ν_e arriving here was found to be a factor of about three smaller than had been expected. This was the first hint that the ν_e might have a mass and be changing into the other varieties of neutrino en-route.

Anomalies were also seen in the mix of ν_e and ν_μ produced when cosmic rays hit atoms in the upper atmosphere. These collisions produce a cascade of secondary particles that eventually lead to a shower of neutrinos at ground level, and even below ground. From experience with the content of cosmic rays and the products of their collisions in the upper atmosphere, it was calculated that there should be twice as many ν_μ as ν_e. However, experiments found a significant shortfall of the ν_μ variety. Neutrinos from the Sun had travelled 150 million km and those in the cosmic ray experiments up to 10,000 km—some having even been produced on the far side of the globe and passed through the centre of the planet before being detected. The data demonstrate the propensity for oscillations en route and the 'disappearance' of ν_e and ν_μ, respectively.

Solar neutrino oscillations have been confirmed by the SNO (Sudbury Neutrino Observatory) in Canada. The experiment is sensitive to all neutrino flavours (see pp. 75–6 in Chapter 6). SNO found that the *total* number of neutrinos arriving here—the electron, muon, and tau varieties—agrees with the number expected based on models of neutrino production in the solar core. SNO also found that the incidence of ν_e is about 1/3 of the total, confirming the shortfall is due to the transformation en route into ν_μ and ν_τ.

SNO showed that ν_e indeed had changed but did not of itself determine into which variety it preferred to go. So have begun 'long baseline' experiments. At accelerators such as CERN, Fermilab, or the KEK laboratory in Japan, controlled beams of neutrinos have been created. The energy, intensity, and composition (mainly ν_μ) of the neutrino beams is monitored at source; it is directed through the ground to be detected several hundred kilometres away at a remote underground laboratory.

By comparing the composition of the arriving beam with that which set out it is becoming possible to determine which flavours oscillate into what, and how quickly they do so. From this it is then possible to calculate information about their relative masses.

Beams of ν_μ, for example, were produced at Fermilab in the MINOS experiment—'Main Injector Neutrino Oscillation Search'—which ended in 2016. The intensity of ν_μ was first measured at source, then again a few hundred metres away, and finally at a large underground detector in Minnesota, 735 km away. In Japan, a neutrino beam created at the KEK laboratory travelled 250 km westwards under the Japanese Alps to the Super-K (Super Kamiokande) detector. In 2010, a tau lepton was seen to have been produced by a beam of ν_μ at CERN, proving that there is a chance for ν_μ to convert to ν_τ. Today the study of neutrino oscillations has become a quantitative science giving information on the mixing for both neutrinos and antineutrinos.

The oscillation data give measures of the relative sizes not of the neutrinos' masses, but the *squares* of them. These are of the order of 10^{-3} to 10^{-4} eV^2. Studies of beta decay and other precision measurements show that neutrino masses are less than 0.8 eV (the mass of the electron, for example, is 511 keV, some 200,000 times larger!). If there were no mixing the three distinct masses of neutrino would match directly one to one for ν_e, ν_μ and ν_τ. The data show that reality is far from this, with the mixing being large, meaning that a neutrino of a given mass has significant affinity for all three flavours. This is summarized in Figure 29.

From the study of the Z^0 boson we know that there are no more light neutrino varieties in Nature. This is because we can measure how long the Z^0 lives, which turns out to be the same as theorists calculated should be the case so long as there are only three distinct varieties of neutrinos that can be produced when it decays. The more varieties there are, the faster the Z^0 would decay as each available path would make the Z^0 more and more unstable.

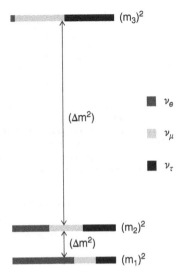

$(m_3)^2$

(Δm^2)

ν_e

ν_μ

ν_τ

$(m_2)^2$

(Δm^2)

$(m_1)^2$

29. **Neutrino 3 flavour mixing. A possible interpretation of data on neutrino mixing. Two of the mass eigenstates are relatively close together while the third is displaced. There is considerable mixing of all three flavours in the two lightest states, while the most massive state is an almost even mixture of ν_μ and ν_τ with a non-negligible presence of ν_e.**

If there are any more light neutrinos, they would shorten the Z^0 life, in disagreement with what is observed in practice. So the inference is that there are only three distinct varieties of such light neutrinos.

Given this result, and as we suspect that every variety of such a neutrino is partnered by a negatively charged lepton, and these in turn partnered by two varieties of quark, the +2/3 and −1/3 variety, then we have identified the full set of such basic pieces. Every one of these leptons and quarks has a spin 1/2. Thus Nature appears to have made three generations of fundamental particles with spin 1/2. Why three? We do not know. Why was it not

Quarks	e = +2/3	u	c	t
	e = −1/3	d	s	b
Leptons	e = −1	e	μ	τ
	e = 0	v_e	v_μ	v_τ

30. Quarks and leptons. The up and down quarks' masses are ~2–5 MeV and the strange ~150 MeV. When trapped inside hadrons they gain extra energy and act as if they have masses of ~350 MeV and ~500 MeV, respectively. The effective masses of the heavier quarks are not so dramatically affected by their entrapment inside hadrons. Their masses are charm ~1.5 GeV, bottom ~4.5 GeV, and top ~180 GeV.

satisfied with one? Here again we do not know for certain, but we suspect that the answer may be related to another puzzle: why is there an imbalance between matter and antimatter in the universe?

Antimatter puzzle

Antimatter has an aura of mystery, the promise of a natural Tweedledum to our Tweedledee, where left is right, north is south, and time runs in reverse. Its most celebrated property is its ability to destroy matter in a flash of light, converting the stuff that we are made of into pure energy. In science fiction, antiplanets tempt travellers to their doom even as antihydrogen powers the engines of astrocruisers. In science fact, according to everything that decades of experimental physics has taught us, the newborn universe was a cauldron of energy where matter and antimatter emerged in perfect balance. Which begs a question: how is it that matter and antimatter did not immediately destroy each other in an orgy of mutual annihilation? How is it that today, some 14,000 million years later, there is anything left in the universe at all?

This conundrum touches on our very existence. We are made of matter, as is everything we know of in the universe. There are no antimatter mines on Earth, which is just as well as they would be destroyed by the matter surrounding them with catastrophic results. Somehow, within moments of the Big Bang, matter had managed to emerge victorious; the antimatter having been annihilated, the heat energy from the destruction remaining (today being a cool 3 degrees above absolute zero in temperature and known as the microwave background radiation), and the surfeit of matter eventually clumping into galaxies of stars. Something must distinguish matter from antimatter so that matter emerged victorious.

The sequence of events that enabled the basic pieces of matter to be cooked within stars, eventually to form bulk matter as we find it today, will be described in the next chapter. Here we discuss the question of how matter and antimatter might differ.

This question has plagued physicists and cosmologists for years. An essential clue turned up in 1964 with the discovery that Nature contains a tiny imbalance, a tendency for the behaviour of certain 'strange' particles, such as the electrically neutral K^o, not to be mimicked precisely by the antimatter counterpart, the $\bar{K}^o$.

The breakthrough came following the discovery in 1977 of the first examples of 'bottom' particles and the realization that they are in effect heavier versions of strange particles. Indeed, when the discovery of bottom and top quarks confirmed that Nature has indeed made three generations of quarks, and of antiquarks, the resulting equations surprisingly seemed to imply that an asymmetry between matter and antimatter for bottom particles was almost inevitable. The subtle asymmetry between K^o and $\bar{K}^o$ was predicted to be rather large for their bottom analogues, the B^o and $\bar{B}^o$. Today we know this to be the case empirically. Could the existence of three generations, and in particular of bottom quarks, somehow hold the key to the conundrum? As bottom particles are abundant in the first moments of the universe, could they hold the secret of how the lopsided universe, where matter dominates today, has emerged?

Today we know that the presence of three generations of quarks or leptons gives mathematically a natural means for particles and antiparticles to behave asymmetrically, and that Nature exploits this. However, it does so most radically in the case of heavy bottom quarks. It turns out these are almost decoupled from the light flavours of which our more stable universe is composed—in the jargon, 'quantum mixing angles are small'. So, we have the tantalizing situation that we have found a mathematically consistent mechanism to generate asymmetry between matter and antimatter, but one that Nature seems not to exploit empirically.

Or at least, not in the case of quarks, for which the mixing angles are small. But in the case of neutrinos, evidence from neutrino oscillations shows that the analogous mixing angles are large. There are also some tantalizing hints that the oscillations of antineutrinos differ subtly from those of their neutrino counterparts (see Chapter 10). Given also that neutrinos are the most abundant particles with mass in the universe, there is speculation that the properties of neutrinos and antineutrinos are somehow key to the matter-antimatter asymmetry. How, or if, this occurs is for the future, but it is stimulating a vast amount of interest in these ghostly particles.

Chapter 9
Where has matter come from?

We exist because of a series of fortunate accidents: the fact that the Sun burns at just the right rate (faster and it would have burned out before intelligent life had the chance to develop; slower and there might not have been enough energy for biochemistry and any life at all); the fact that protons—the seeds of hydrogen—are stable, which enables stars to cook the chemical elements essential for the Earth to be built; and the fact that neutrons are slightly heavier than protons, which enables beta radioactivity, transmutation of the elements such as the protons of hydrogen into helium, which in turn enables the Sun to shine. Were any of these, or several others, slightly changed, we would not be here.

We and everything are made from atoms. Where did these atoms come from? Most recently (by which I mean 5 billion years!) they were formed inside a long-dead star where they were all cooked from protons, the nuclei of the simplest atomic element—hydrogen. The protons were formed very early in the universe and its constituent quarks, and also the electrons, were made within the first moments. This chapter describes how the stuff that we are made of came to be.

It is primarily protons that form the Sun and fuel it today. Let's first describe how the Sun works and provides the energy for us to exist. Hydrogen is the simplest atom, where a single negatively charged electron encircles a central positive proton. Hydrogen may be relatively uncommon on Earth (except when trapped inside molecules such as water—H_2O), but in the universe at large it is the most common atomic element of all. At Earthly temperatures atoms can survive, but at higher temperatures, above a few thousand degrees, the electrons are no longer trapped but roam free: the atom is said to be ionized. This is what it is like inside the Sun: electrons and protons swarm independently in the state of matter known as plasma.

Protons can bump into one another and initiate a set of nuclear processes that eventually converts four of them into the nuclei of the next simplest element: helium. The energy locked into a single nucleus of helium (its $E = mc^2$) is less than that in the original four protons. This 'spare' energy is released into the surroundings, some of it eventually providing warmth here on Earth.

The protons have to touch in order to fuse and build up helium. This is hard as their positive charges tend to repel them, keeping them apart. However, the temperature of 10 million degrees gives them enough kinetic energy that they manage to encroach near enough to start the fusion power process. But it is only just enough: 5 billion years after its birth, any individual proton has only a 50:50 chance of having taken part in the fusion. Put another way: thus far the Sun has used up half of its fuel.

This is the first fortunate circumstance. Humans are the pinnacle of evolution, and it has taken almost all of those 5 billion years for us to emerge. Had the Sun burned faster, it would have died before we arrived.

So let's see what happens and then why it is balanced just right.

The first step is when two protons meet and touch. One of them undergoes a form of radioactive decay, turning into a neutron and emitting a positron (the antiparticle of an electron) and a neutrino. Normally it is the neutron that decays, due to its extra mass and associated instability, into a proton, electron, and neutrino. An isolated proton being the lightest baryon, by contrast, is stable. But when two protons encroach, they feel electrostatic repulsion; this contributes to their total energy, making it exceed that of a deuteron (a proton and neutron bound together). As a result, one of the protons can turn into a neutron, which then binds to another proton, increasing the stability. This decay of the proton leads to a neutron, neutrino, and positron, the positively charged antiparticle of an electron.

So the very first part of the solar fusion cycle produces antimatter! The positron is almost immediately destroyed as it collides with an electron in the plasma, producing two photons which are scattered by the electrically charged plasma, eventually working their way to the solar surface (this takes several thousand years), by which time their energy is much reduced and they help form part of sunlight. The neutrinos pour out from the centre unhindered and reach us within a few minutes.

So what has become of the neutron and proton? They grip one another tightly, courtesy of the strong nuclear force, and bind together: this doublet is a nucleus of heavy hydrogen—the deuteron. This deuteron finds itself in the midst of a vast number of protons, which still form the bulk of the Sun. Very rapidly the deuteron links with another proton to make a nucleus of helium: helium-3. Two of these helium-3 can join and rearrange their pieces to form a nucleus of helium-4 (the stable common form), releasing two spare protons.

So the net result of all this is that four protons have produced a single helium, two positrons, and two neutrinos. Protons are the fuel, helium the ash, and the energy is released in the form of gamma rays, positrons, and neutrinos.

The latter steps, where a deuteron and a proton make ^{3}He and then lead to ^{4}He, happen almost instantaneously; it is the tardiness of the first step, $p + p \rightarrow d v e^+$ that controls the (slow) burning of the Sun that has been so important for us.

The rate of the burning depends upon the strength of the weak force, which transmutes the proton into a neutron ('inverse beta decay'). This force has parallels with the electromagnetic force, as described earlier. The electromagnetic force is transmitted by photons, which are exchanged between one electrically charged

At the heart of the Sun:

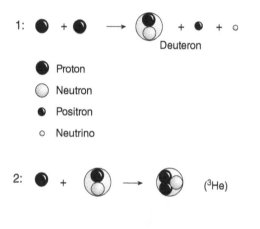

31. **Converting hydrogen to helium in the Sun.**

particle and another. Photons are massless: this enables them to spread to large distances without restrictions from energy conservation and hence gives the electromagnetic force a long range. The weak force, by contrast, owes its feebleness (at least at the energies characteristic of Earth and the Sun) to the large mass of the W boson and its consequent restricted range.

The slowness of solar burning is controlled by the feebleness of the weak force which is in turn controlled by the large mass of the W boson. Had its mass been smaller, the effective strength of the 'weak' force would have been stronger and the rate of solar burning faster. The W gains its mass by its interactions with the Higgs field. So here we see a key consequence of these dynamics: the Sun burning slowly has enabled long enough to elapse for evolution of intelligent life to occur. We owe our existence to the effects of the Higgs field.

There are other examples where masses play a sensitive role in determining our fate. As we have discussed above, beta decay involves a neutron turning into a proton and emitting an electron and a neutrino. This requires the neutron to be heavier than the proton—which it is, whereby protons are the stable seeds of atoms and chemistry. (Had neutrons been lighter, then it would have been neutrons that emerged as the stable pieces from the Big Bang. These neutral particles would have been unable to attract electrons to form atoms, so chemistry would have been different, or non-existent.) The neutron is only one part in a thousand heavier than the proton, but fortunately this is enough that an electron can be produced, or put another way, the electron mass is small enough that it can be produced in such a process. Had it been larger, then beta decay and the Sun would have been frozen; had it been smaller, beta decay would have been faster, the Sun's dynamics different, the intensity of ultraviolet light higher and unhealthy for us. (The mass of the electron helps determine the size of atoms such as hydrogen; smaller mass correlates with a larger atom and vice versa. So things have the size they do in part because the mass of the electron is as it is.) The reason for this pattern of masses is still to be found.

So the Sun is shining courtesy of nuclear fusion. In another 5 billion years its hydrogen will all have gone, turned into helium. Already some of the helium is itself fusing with protons and other helium nuclei to build up the nuclear seeds of heavier elements. These processes also produce neutrinos, some of higher energies than those produced in the primary proton fusion; and so by detecting neutrinos from the Sun, and measuring their energy spectrum, we can begin to get a quantitative look inside our nearest star.

Five billion years hence these will be the primary processes, along with the fusion to build up yet heavier elements. In some stars (but not our Sun) this process continues, building up the nuclei of elements up to iron, which is the most stable of all (there are even elements beyond iron that are built, but they tend to be rarer). Eventually such a star is unable to resist its own weight, and it collapses catastrophically. The shock waves spew out matter and radiation into space. This is known as a supernova. So stars begin as hydrogen, and with these ingredients they cook the periodic table; a supernova is the agent that pollutes the cosmos with the nuclear seeds of these chemicals.

So where did the material for the primary stars come from?

The early universe

The basic pieces of nuclear matter, quarks, emerged from the Big Bang along with electrons. The universe rapidly cooled so that the quarks clustered together to form protons. The following processes took place:

$$e(electron) + p(proton) \rightleftharpoons n(neutron) + \nu(neutrino)$$

The double arrow is to illustrate that this process could occur in either direction. The neutron is slightly heavier than the combined masses of a proton and an electron, so the 'natural' direction for the processes was to go from right to left: the neutron has a natural tendency to lower the mass of the whole, liberating energy via $E = mc^2$. However, the heat of the universe was such that the electrons and protons had considerable amounts of kinetic energy, such that their total energy exceeded that locked into the mass (mc^2) of a neutron. So in these hot conditions the process could as easily run from left to right (electron and proton converting into neutron and neutrino) as the other direction where the neutrons and neutrinos turned back into their electrically charged cousins. In these circumstances we say that the universe was in thermal equilibrium.

But the universe was rapidly cooling, which made it harder for the production of neutrons to continue. After a microsecond the universe had cooled to a point where this neutron production reaction was effectively frozen out. The surviving reaction was

$$n \rightarrow p + e + \bar{\nu}$$

During this epoch any neutrons that had been produced in the earlier heat would be dying out. Every 10 minutes their numbers halved (we say they have a 'half-life' of about 10 minutes). There was no longer enough energy to replace them. But not all the neutrons died, as some fortunate ones bumped into protons, whereupon they fused to one another to make a deuteron (a bound system of a single proton and a neutron which is lighter than an isolated proton and neutron are).

At this stage the universe at large plays out the sequence that is going on in the Sun today: deuterons and protons building up nuclei of helium. This took place until either all of the neutrons had died out and gone forever, or the particles in the expanding universe were so far apart that they no longer interacted with one another.

One microsecond after the Big Bang, all of the neutrinos produced in these reactions were free. They thus became the first fossil relics of the universe. They moved at high speed and their mass, although very small, gave enough gravitational attraction among the hordes that they started clustering together, contributing to the formation of galaxies. About a billion neutrinos are produced for every atom that eventually forms. Neutrinos are thus among the most populous particles in the universe. Although we know that at least one of the varieties of neutrinos has mass, we don't yet know how big this is. If the mass of a neutrino is greater than a few eV, that is, a billionth of that of a proton, then neutrino masses will dominate the mass density of the material universe. So determining the mass of neutrinos can be a big issue for predicting the long-term future of the universe. Will it expand forever, or eventually collapse under its own weight? We don't yet know for sure.

The universe continues to expand and cool. The principles of physics that determine its expansion are in some ways similar to those that control the behaviour of a gas in a container. The rate depends on the pressure, which depends on the temperature in the gas and the number of neutrinos inside the gas volume (the density). This in turn depends on the number of neutrino species.

Three minutes after the Big Bang, the material universe consisted primarily of the following: 75 per cent protons; 24 per cent helium nuclei; a small amount of deuterons; traces of other light elements; and free electrons.

The abundance of helium and of the light elements depends on the expansion rate of the universe, which in turn depends on the number of neutrino species. The observed amount of helium fits with predictions if there are three varieties of neutrino. The fact that measurements of the Z boson at CERN showed that there are indeed three varieties of light neutrino is a remarkable agreement between measurements in particle physics, which replicates the conditions of the early universe, and what cosmologists had inferred from the above.

The abundance of deuterium depends on the density of 'ordinary' matter in the universe (by ordinary we mean made of neutrons and protons, as against other exotic things that theorists might dream of but for which there is as yet no direct experimental proof, for example supersymmetry, see Chapter 10). The numbers all fit provided that the density of ordinary matter is much less than the total in the universe. This is part of the dark matter puzzle: there is stuff out there that does not shine but is felt by its gravity tugging the stars and galaxies. It seems that much of this must consist of exotic matter whose identity is yet to be determined.

Some 300,000 years later, the ambient temperature had fallen below 10,000 degrees, that is similar or cooler than the outer regions of our Sun today. At these energies the negatively charged electrons were at last able to be held fast by the electrical attraction to the positively charged atomic nuclei whereby they combined to form neutral atoms. Electromagnetic radiation was set free and the universe became transparent as light could roam unhindered across space.

The universe has expanded and cooled for almost 14 billion years so far. The once hot electromagnetic radiation now forms a black body spectrum with an effective temperature of about 3 degrees above absolute zero. The discovery of this by Penzias and Wilson half a century ago is one of the great pieces of support for the Big Bang theory. Today precision measurements of the spectrum by instruments in satellites reveal small fluctuations in the cosmic microwave. These give hints of proto galaxies forming in the early universe.

So we have a good qualitative and even quantitative understanding of how the basic seeds of matter ended up in you and me. But as they emerged along with antimatter in that original Big Bang, a puzzle remains: where did all the antimatter go? That is one of the questions whose answer is still awaited.

Chapter 10
Questions for the post-Higgs era

The Higgs field

The electroweak force is the force carried by the familiar photon of electromagnetism, and by the W and Z bosons, which are responsible for the weak interactions that not only initiate solar burning, but also underlie certain types of radioactivity. Yet if these effects are so closely intertwined, why do they appear so different in our daily experiences, that is, at relatively low temperatures and energies? One reason is that the particle that transmits the electromagnetic force, the photon, is massless, whereas the W and Z bosons, which are associated with the weak force, have huge masses and each 'weighs' as much as an atom of silver.

The standard model of the fundamental particles and the forces that act among them explains mass by proposing that it is due to a field named after Peter Higgs who in 1964 was one of the first to recognize this theoretical possibility. The Higgs field permeates all of space. Were there no Higgs field, according to the theory, the fundamental particles would have no mass. What we recognize as their mass is the effect of the interaction between particles and the Higgs field. Photons do not interact with the Higgs field and so are massless. The data confirm that the mass of W and Z bosons comes from the Higgs mechanism, as do the masses of top, bottom, and charm quarks also. It is too early to determine whether the mass of lighter quarks, and of the leptons also, arises from this mechanism, though data for muon and tau leptons are consistent with it. This will become answered as data accumulate in the coming years. The masses of protons and neutrons, however, are dominated by the kinetic energy of their constituent quarks and are only marginally related to the Higgs field.

Just as electromagnetic fields produce the quantum bundles we call photons, so does the Higgs field manifest itself in Higgs bosons. In Higgs' original theory there was just one type of Higgs boson, but if supersymmetry is correct (see below), there should be a family of such particles.

With the discovery of the Higgs boson, the picture of particle physics is now internally complete, in the sense that it is mathematically consistent. Six quarks matched by six leptons, grouped in three pairs of flavours, interact by the exchange of vector bosons, their masses coming from interactions with the Higgs field. The resulting quantum field theories of these particles and forces describe data to superb accuracy.

Successful though this theory may be, we know it doesn't fully describe our world. Internal completeness is a mathematical requirement, whereas describing the world around us is the demand of natural philosophy. Even apart from the lack of a quantum theory of gravity, the standard model doesn't describe dark matter (see below)—which on a cosmic scale far outweighs the stuff that we identified and had begun to explain in the pre-Higgs era. The Higgs mechanism explains how masses arise but gives no clue why they are as they are—why is the up quark lighter than a down quark, making the proton lighter than a neutron, for example? There is no explanation for the number of fundamental leptons and quarks, nor an answer even to a question as basic as 'What makes an electron an electron?' What, indeed, is flavour? Why does the weak interaction violate

124

parity—mirror symmetry? Why does the Higgs field act in such a discriminatory way, leaving the photon massless while giving huge masses to the W and Z bosons? Other than an anthropic answer—we would not exist were it not like this—we have no fundamental explanation of this asymmetry.

Is the Higgs field alone the source of the mass of leptons, of quarks, and the W and Z bosons, or is there some other contribution, and, if so, might identifying the latter explain the nature of flavour? This is an example of a question that requires high-precision data on the decays of Higgs bosons in hope of finding some systematic deviation from expectation that might expose subtle phenomena.

Nor do we know whether the gravitational, electroweak, and strong forces are the totality. It is straightforward to generalize the mathematical structures that underpin these gauge theories leading to the prediction that more forces exist, but there is no evidence that Nature makes use of this. Now that the mass mechanism has been established, it is plausible that the gauge bosons of these generalizations are so massive that the resulting forces are too feeble for us yet to discern. If, however, gravity, the electroweak, and strong forces are the totality, we are left with a question: why?

Studying the Higgs boson with high precision might reveal some deviations from theoretical expectations and thereby point the way to further progress. In the immediate future we need to understand how Higgs bosons mutually interact, ultimately condensing to form the Higgs field. By analogy, we need the Higgs field much like fish need water, and our discovery of the Higgs boson is akin to fish having discovered a molecule of H_2O when the goal is to understand the ocean. At the LHC in the coming decade, higher intensity of the beams and potentially higher energy too will increase the chances of producing two (or more) Higgs bosons in a single event. From this we can study how they

32. Peter Higgs at the CMS detector.

mutually interact and begin to understand how the Higgs field is formed. Is it a featureless basic field or itself made of deeper structure? We know the Higgs field is there; its dynamic structure and even what it is made of remain a mystery.

Dark matter

Protons and the nuclei of ordinary atoms seed all the 'luminous matter' that shows up in astronomical observations. However, the motions of spiral galaxies, to take one example, show that there is more gravitational force than the observed luminous matter can account for. As much as 90 per cent of the matter present remains undetected. It appears that the universe we see by its electromagnetic radiations is outweighed by some mysterious 'dark matter', which does not show up at any wavelength in our telescopes.

If there are large 'massive compact halo objects' (MACHOs) which could be bodies about the size of Jupiter, and not big enough to become shining stars or black holes, they would be detectable by creating double or multiple images of the distant star or galaxy through the effect of gravitational lensing. However, searches of this kind have not found enough MACHOs to explain the vast amount of dark matter that the universe appears to contain. The detection of gravitational waves arising from the collisions of black holes is giving information on their abundance and masses. These too seem unable to account for the amount and distribution of dark matter. Astrophysicists and cosmologists have had to turn to particle physics for further ideas.

The intriguing possibility is that this dark matter could consist of vast quantities of subatomic particles that do not interact electromagnetically (otherwise we would be able to detect their electromagnetic radiation). One obvious candidate is the neutrino, whose tiny but non-zero mass could cause large clouds of them to gravitate to one another and help seed the formation of the galaxies.

In the early universe, these neutrinos would have been highly energetic, moving at almost the speed of light. In the jargon, such flighty entities are known as 'hot', and computer simulation of galaxy evolution in a 'hot dark matter' universe shows galaxies forming in dense clusters with large voids between them. However, this computer model of the universe does not look like what the astronomers observe in practice.

The evolution of galaxies would have been very different if the dark matter consists of massive, slow-moving, and therefore 'cold', particles. A problem is that there are no such entities known in the standard model, so if this is the answer to the dark matter problem, it raises another question: who are these particles?

The existence of dark matter may be the first clue to what lies beyond the standard model. A favoured theory postulates the existence of 'supersymmetric' particles, the lightest of which include forms that do not respond to the electromagnetic or strong forces, but which may be hundreds of times more massive than the proton. Collisions at the highest-energy particle accelerators, such as the Tevatron at Fermilab and the Large Hadron Collider at CERN, have yet to find evidence for such particles, however. A problem is that we have no reliable guide on how massive they might be and whether they are within reach at the LHC. Cosmic rays might produce such 'dark particles', but here too we as yet have no empirical confirmation of them.

Which brings us to the question: what is supersymmetry?

Supersymmetry

The paradox of how 'empty' atoms form solid matter is solved in quantum mechanics. It is a profound property of the fact that electrons (and quarks, and protons, and neutrons) all have an intrinsic spin that is one-half of an amount known as Planck's constant, h. Such 'spin 1/2' particles are generically known as fermions. Quantum mechanics implies that two fermions cannot be in the same place, with the same state of motion; in the jargon they 'cannot occupy the same quantum state'. This causes the many electrons in complex atoms to occupy specific states, and gives rise to the chemical activity, or inertness, of the various elements. It also prevents an electron in one atom encroaching too readily on one in a neighbouring atom. This underpins many properties of bulk matter, such as solidity.

The forces among these fermions are transmitted by photons, gluons, and W and Z bosons. Note the word 'bosons'—the generic term referring to particles that have a spin that is zero or an integer multiple of Planck's quantum. Recall that in contrast to fermions, which are mutually exclusive, bosons have affinity and form collective states, such as is the case for photons in laser beams. One of the goals at the LHC is to produce pairs of Higgs bosons at a time. By studying how they mutually interact we may be able to determine how Higgs bosons collectively form the Higgs field, in which we are immersed.

We have seen that the fermions—quarks and leptons—exhibit a profound unity, and also that the force-carrying bosons do too. Why is it that 'matter particles' are all (apparently) made of spin-1/2 fermions and the forces transmitted by spin-1 bosons? Could there be a further symmetry between the forces and the matter particles, such that the known fermions are partnered by new bosons, and the known bosons by new fermions, with novel forces transmitted by these fermions? Could this lead to a more complete unification among particles and forces? According to the theory known as supersymmetry, the answer is yes.

In supersymmetry—or SUSY as it is known—there are families of bosons that twin the known quarks and leptons. These

129

'superquarks' are known as **squarks**; their superlepton counterparts are known as **sleptons**. If SUSY were an exact symmetry, each variety of lepton or quark would have the same mass as its slepton or squark sibling. The electron and selectron would have the same mass as one another; similarly, the up quarks and the 'sup' squark would weigh the same, and so on. In reality this is not how things are. The selectron, if it exists, has mass far greater than 100 GeV, which implies that it would be hundreds of thousands of times more massive than the electron. Similar remarks can be made for all of the sleptons or squarks.

An analogous statement can also be made about the super-partners of the known bosons. In SUSY there are families of fermions that twin the known bosons. The naming pattern here is to add the appendage '-ino' to denote the super-fermion partner of a standard boson. Thus there should exist the photino, gluino, zino, and wino (the 'ino' pronounced eeno, thus for example it is weeno and not whine-o). The hypothetical graviton, the carrier of gravity, is predicted to have a partner, the gravitino. Here again, were supersymmetry perfect, the photino, gluino, and gravitino would be massless, like their photon, gluon, and graviton siblings; the wino and zino having masses of 80 and 90 GeV like the W and Z. But as was the case above, here again the 'inos' have masses far greater than their conventional counterparts. If SUSY is correct, there should be one or more fermion analogues of the Higgs boson, so called 'Higgsinos'.

SUSY may be theoretically seductive, but we have not found clear evidence for a single squark or slepton, nor photino, gluino, wino, zino, or Higgsino. Searching for them at the LHC and elsewhere is a high priority.

With such a lack of evidence for superparticles, one might wonder why theorists believe in SUSY at all. It turns out that such a symmetry is very natural, at least mathematically, given the nature of space and time as encoded in Einstein's theory of relativity and

quark	q ;	squark	$\tilde{q}$
leptons	l ;	sleptons	$\tilde{l}$
electron	e ;	selectron	$\tilde{e}$
neutrino	ν ;	sneutrino	$\tilde{\nu}$
photon	γ ;	photino	$\tilde{\gamma}$
gluon	g ;	gluino	$\tilde{g}$
W boson	W ;	Wino	$\tilde{W}$
Z boson	Z ;	Zino	$\tilde{Z}$
Higgs Boson	H ;	Higgsino	$\tilde{H}$

Some particles and their super-partners

33. **Supersymmetry particles summary: massive neutrinos and oscillations.**

the nature of quantum theory. The resulting pattern of superparticles turns out to solve some technical problems in the present formulation of particle physics, stabilizing the quantum theories of the behaviour of the different forces at high energies and the responses of particles to those forces. In a nutshell, without SUSY certain attempts to construct unified theories lead to nonsensical results, such as that certain events could occur with an infinite probability. However, quantum fluctuations, where particles and antiparticles can fleetingly emerge from the vacuum before disappearing again, can be sensitive to the SUSY particles as well as to the known menu; upon including the SUSY contributions, more sensible results emerge. The fact that the nonsensical results have disappeared encourages hope that SUSY is indeed involved in Nature's scheme. Getting rid of nonsense is, of course, necessary, but we still do not know if the sensible results are identical with how Nature actually behaves. So we have at best indirect hints that SUSY is at work, albeit behind the scenes at present. The challenge is to produce SUSY particles in experiments, thereby proving the theory and enabling detailed understanding of it to emerge from the study of their properties.

SUSY might be responsible for at least some of the dark matter that seems to dominate the material universe. From the motions of the galaxies and other measurements of the cosmos, it can be inferred that perhaps as much as 90 per cent of the universe consists of massive 'dark' matter, dark in the sense that it does not shine, possibly because it is impervious to the electromagnetic force. In SUSY if the lightest superparticles are electrically neutral, such as the photino or gluino say, they could be metastable. As such they could form large-scale clusters under their mutual gravitational attraction, analogous to the way that the familiar stars are initially formed. However, whereas stars made of conventional particles, and experiencing all the four forces, can undergo fusion and emit light, the neutral SUSY-inos would not. If SUSY particles are discovered, it will be fascinating to learn if the required neutral particles are indeed the lightest and have the required properties. If this should turn out to be so, then one will have a most beautiful convergence between the field of high-energy particle physics and that of the universe at large.

Quark gluon plasma

If our picture of the origins of matter is correct, then the quarks and gluons, which in today's cold universe are trapped inside protons and neutrons, would in the heat of the Big Bang have been too hot to stick together. Instead, they would have existed in a dense, energetic 'soup' known as 'quark gluon plasma', or QGP for short.

These intermingled swarms of quarks and gluons are analogous to the state of matter known as plasma, such as is found in the heart of the Sun, which consists of independent gases of electrons and nuclei too energetic to bind together to form neutral atoms.

Physicists make QGP by smashing large atomic nuclei into one another at high energies, such as at the LHC. In such extreme conditions the protons and neutrons squeeze together. The nuclei 'melt'—in other words, the quarks and gluons flow throughout the nucleus rather than remaining 'frozen' into individual neutrons and protons.

The Relativistic Heavy Ion Collider (RHIC) at Brookhaven National Laboratory in the USA is a dedicated machine where beams of heavy nuclei collide head on. As with simpler particles, such as electrons and protons, the great advantage of a colliding beam machine is that all the energy gained in accelerating the particles goes into the collision. RHIC has been superseded in energy by the Large Hadron Collider at CERN, where lead ions collide at a total energy of up to 1,300 TeV. At these extreme energies, akin to those that would have been the norm in the universe when it was less than a trillionth of a second old, QGP should become commonplace, so that experimenters can study its properties in detail.

Massive neutrinos and the antimatter mystery

A wealth of information about the enigmatic neutrinos is coming from 'long baseline' experiments. Determining the pattern of their masses and their propensity to oscillate into one another will provide some of the missing parameters of the standard model. We do not know why the values of the masses of the quarks and charged leptons are as they are. That they have those values is critical for our existence, so understanding this would be a significant breakthrough. Determining the neutrino masses could therefore provide an essential clue in unravelling this enigma.

There remains the possibility that there are more massive neutrinos, separate from the trio of light ones. 'Sterile' neutrinos are hypothetical neutral leptons that interact only by gravity and not by other forces in the standard model. If sterile neutrinos exist and their mass is smaller than the energies of particles in the experiment, they can be produced in the laboratory, either by quantum mixing between conventional 'active' neutrinos and sterile neutrinos or in high-energy particle collisions. In 2018 the MiniBOONE experiment at Fermilab reported a stronger neutrino oscillation signal than expected, which raised excitement that this might be a hint of sterile neutrinos playing some role. However, in 2021, results of an upgraded experiment, known as MicroBOONE, showed no evidence of sterile neutrinos. This remains an active area of research.

Neutrino masses could also have impact on cosmology. Massive neutrinos could have played a role in seeding the formation of galaxies; they could play some role in explaining the nature of the dark matter that pervades the universe; and there is still the unresolved puzzle of why the weak interaction experiences a violation of parity, mirror symmetry. Neutrinos are a special entree to probing the weak interaction, and so the increased study of their properties may lead to unexpected discoveries.

There is also theoretical speculation that neutrinos might have played a key role in generating the large-scale asymmetry between matter and antimatter in the universe. This enigma has grown ever more tantalizing since the original discovery of CP violation—in effect a difference between matter and antimatter—in the strange particles in 1964. It seems that we inhabit a volume of matter that is at least 120 million light years in diameter. Based on the behaviour of strange particles, and more recently of their bottom analogues, most physicists favour the idea that there is some subtle asymmetry between matter and antimatter at large, and that soon after the Big Bang this tipped the balance in favour of a universe dominated by matter.

The challenge now is to study these differences in detail in order to identify their origins and, perhaps, the source of the asymmetry between matter and antimatter in the cosmos.

Recall, kaons are made of a quark and an antiquark and as such are an equal mixture of matter and antimatter. The neutral kaon (K^0) consists of a down quark and a strange antiquark, while its antiparticle consists of a down antiquark and a strange quark. The K^0 and $\bar{K}^0$ are thus different particles, but they are intimately related through the weak force which, rather surprisingly, allows a K^0 to change to a $\bar{K}^0$, and vice versa, via interactions between their quarks and antiquarks. What this effect means is that once a neutral kaon or neutral antikaon is created, some quantum mechanical 'mixing' begins to occur.

These in-between mixtures are known as the K_S (S for 'short') and the K_L (L for 'long'). The K-Long lives about 600 times longer than the K-Short. The important feature is that the states K-Long and K-Short behave differently in the combined 'mirrors' of CP. The two states decay in different ways, the K-Short to two pions, the K-Long to three pions. If CP symmetry were perfect this pattern of decay would always be true. The K-Long, for example, would never decay to two pions. However, as Cronin and Fitch and their colleagues first observed in 1964, in about 0.3 per cent of cases the K-Long does decay to two pions.

With the discovery of bottom quarks in the late 1970s, theorists realized that the presence of three generations can lead naturally to an asymmetry between matter and antimatter, as outlined in Chapter 8. Theory predicted that CP violation should be a large effect in the case of B mesons, which are similar to kaons but with the strange quark replaced by a bottom quark. This has been confirmed by experiments at customized 'B-factories' and by the LHCb detector at CERN. However, this does not seem to explain the vast asymmetry observed for bulk matter made of up and down quarks.

The discovery that there are also three flavours of massive neutrinos is leading theorists to investigate whether three generations of leptons could have seeded the empirical asymmetry. This has gained credence following the discovery that neutrinos appear more likely to oscillate than antineutrinos and that mixing is more pronounced than is the case for the quarks, as mentioned in Chapter 8.

Since 2010, mu-neutrinos and mu-antineutrinos produced in Tokai, Japan, have been beamed 295 km to Kamioka, the underground neutrino observatory. This is known as the 'T2K' experiment. A possible difference in oscillation rates for neutrinos and antineutrinos was first noticed in 2016. By 2020 the significance had grown. If correct, the data suggest that neutrino mixing is as large as theoretically possible. More sensitive experiments are being designed to verify if this is the case.

If confirmed, this would be an interesting discovery in its own right. More exciting is the possibility that it could be linked to the large-scale matter asymmetry of the universe. This would follow if there are super-heavy neutrinos that partner the currently known lightweight trio, and if these too exhibit an analogous asymmetry to their lighter cousins. This is why.

If there are heavy neutrinos, their decays will spawn lightweight quarks and antiquarks of all flavours. These in turn cascade down to leave the stable seeds of conventional matter and, potentially, antimatter. However, if the massive neutrals exhibit an analogous asymmetry to their lightweight cousins, their decays will lead to an asymmetry in the numbers of quarks and antiquarks. Thus, it is possible that we have identified a route to understanding the antimatter puzzle. It requires establishing an asymmetry between neutrinos and antineutrinos, finding massive neutral cousins to these fermions, and establishing that they too exhibit an asymmetry. Whether any of this is true, hopefully time will tell. Establishing a difference between the oscillation rates of neutrinos and antineutrinos should be possible within a few years. Discovering massive neutrinos, if they exist, could happen at any time, or not at all. If they are found this will be a major news story, so if there is nothing about them, you can assume they are still unseen. Once established, attention will focus on studying their decay properties to see if matter is favoured over antimatter. That could take many years and become a major focus of particle physics research.

Science fiction or science fact?

Now for the really bizarre. According to some attempts to build a quantum theory of gravity, the three dimensions of space and that of time may be just a part of a more profound universe. If so, there are dimensions that are beyond perception by our usual senses, but which could be revealed at the LHC.

To make some sense of this, imagine a universe perceived by flatlanders who are aware of only two dimensions. We, with our greater awareness, know of a third. So we can imagine two flat plates separated by, say, a millimetre. The effects of forces that act within the two dimensions of one plate could leak into the third dimension. Some fraction of this leakage will cross the gap and be experienced by the flatlanders living on the adjacent sheet. They would perceive only the remnant effects, which would be feeble in comparison to the fundamental effects that would be experienced when not restricted to the flat plane universe that they experience.

Now imagine us as 'flatlanders' in a universe with higher dimensions. The idea is that gravity appears feeble to us because it is the effect of the other forces leaking out into the higher dimensions in our universe. So when we feel gravity, we are feeling the effect of the other unified forces that have leaked away into the higher dimensions, leaving a trifling remnant to do its work. One could even imagine particles moving from our 'flatlander' dimensions into the higher dimensions and in effect 'disappearing' from the universe as we know it.

In experiments at the LHC at CERN, physicists are on the lookout for signs of particles 'spontaneously' appearing or vanishing. If such a phenomenon were found to occur in some systematic way, this could provide evidence that we are indeed like flatlanders, and that there are dimensions in Nature beyond the three space and one time that we currently experience.

We have reached a point where it is beginning to be hard to distinguish science fact from science fiction. But a century ago, much of what we take for granted today would have been beyond the imagination of H. G. Wells. A hundred years from now there will be material in the science text books as yet undreamed of. Some fifty years ago I read a book that told of the wonders of the atom as they were then being revealed, and of the strange particles that were showing up in cosmic rays. Today I am writing about them for you. Perhaps in another half century some of you may be updating the story for yourselves. Good luck.

Further reading

The following suggestions for further reading are not intended to form a comprehensive guide to the literature on particle physics.

This section includes some 'classics' that are out of print, but which should be available through good libraries or second-hand bookshops, on the ground or via the Internet (such as at www.abebooks.com).

Sean Carroll (2013) *The Particle at the End of the Universe*. OneWorld. A popular pedagogic account of the standard model and the role of the Higgs boson.

Frank Close (2000) *Lucifer's Legacy*. Oxford University Press. An interesting introduction to the meaning of asymmetry in particle physics and Nature.

Frank Close (2007) *The New Cosmic Onion: Quarks and the Nature of the Universe*. Taylor and Francis. An account of particle physics in the 20th century for the general reader.

Frank Close (2012) *Neutrino*. Oxford University Press. The story of the neutrino, focusing on Ray Davis' quest to find solar neutrinos.

Frank Close (2015) *Nuclear Physics—A Very Short Introduction*. Oxford University Press. Companion volume to this but focusing on nuclear physics, including some aspects of particle physics.

Frank Close (2018) *Antimatter*. Oxford University Press. Antimatter's meaning, discovery, and application, together with a compendium of facts and fictions.

Frank Close (2022) *Elusive*. Allen Lane. A biography of Peter Higgs and the quest for the elusive Higgs boson.

Frank Close, Michael Marten, and Christine Sutton (2003) *The Particle Odyssey*. Oxford University Press. A highly illustrated popular journey through nuclear and particle physics of the 20th century, with pictures of particle trails, experiments, and the scientists.

Graham Farmelo (2009) *The Strangest Man*. Faber and Faber. Award-winning biography of Paul Dirac, complete with a pedagogic introduction to quantum electrodynamics.

Gordon Fraser (ed.) (1998) *The Particle Century*. Institute of Physics. The progress of particle physics through the 20th century.

Brian Greene (1999) *The Elegant Universe: Superstrings, Hidden Dimensions, and the Quest for the Ultimate Theory*. Jonathan Cape. A prize-winning introduction to the 'superstrings' of modern theoretical particle physics.

Tony Hey and Patrick Walters (1987) *The Quantum Universe*. Cambridge University Press. An introduction to particle physics and quantum theory.

George Johnson (2000) *Strange Beauty: Murray Gell-Mann and the Revolution in Twentieth-century Physics*. Jonathan Cape. A biography of Murray Gell-Mann, the 'father' of quarks.

Gordon Kane (1996) *The Particle Garden: Our Universe as Understood by Particle Physicists*. Perseus Books. An introduction to particle physics and a look at where it is heading.

Robert Weber (1980) *Pioneers of Science*. Institute of Physics. Brief biographies of physics Nobel Prize winners from 1901 to 1979.

Steven Weinberg (1993) *Dreams of a Final Theory*. Pantheon Books, 1992; A 'classic' on modern ideas in theoretical particle physics.

Steven Weinberg (1993) *The First Three Minutes*. Andre Deutsch, 1977; Basic Books. The first three minutes after the Big Bang, described in non-technical detail by one of the 20th century's leading theorists.

Frank Wilczek (2015) *A Beautiful Question*. Allen Lane. The standard model described by one of its architects.

W. S. C. Williams (1994) *Nuclear and Particle Physics*, revised edn. Oxford University Press. A detailed first technical introduction suitable for undergraduates studying physics.

Glossary

alpha particle: two protons and two neutrons tightly bound together; emitted in some nuclear transmutations; nucleus of a helium atom.

angular momentum: a property of rotary motion analogous to the more familiar concept of momentum in linear motion.

antimatter: for every variety of particle there exists an antiparticle with opposite properties such as the sign of electrical charge. When particle and antiparticle meet, they can mutually annihilate and produce energy.

anti(particle): antimatter version of a particle, for example antiquark, antiproton.

ATLAS: 'A Toroidal LHC ApparatuS'—detector at the LHC.

atom: system of *electrons* encircling a nucleus; smallest piece of an element that can still be identified as that element.

B: symbol for the 'bottom meson'.

B-factory: accelerator designed to produce large numbers of particles containing *bottom quarks* or antiquarks.

baryon: class of *hadron*; made of three *quarks*.

beta decay (beta radioactivity): nuclear or particle transmutation caused by the *weak force*, resulting in the emission of a *neutrino* and an *electron* or *positron*.

boson: generic name for particles with integer amount of *spin*, measured in units of *Planck's constant*; examples include carriers of forces, such as *photon, gluon, W* and *Z bosons*, and the (predicted) spinless *Higgs boson*.

bottom(ness): property of *hadrons* containing *bottom quarks* or antiquarks.

bottom quark: most massive example of *quark* with electric charge –1/3.

bubble chamber: form of particle detector, now obsolete, revealing the flightpaths of electrically charged particles by trails of bubbles.

CERN: European Centre for Particle Physics, Geneva, Switzerland. The acronym CERN is derived from the French for 'European Centre for Nuclear Research'.

charm quark: *quark* with electric charge +2/3; heavy version of the *up quark* but lighter than the *top quark*.

CMS: Compact Magnetic Solenoid detector at the LHC.

collider: particle accelerator in which beams of particles moving in opposing directions meet head on.

colour: whimsical name given to property of *quarks* that is the source of the *strong forces* in the *QCD* theory.

conservation: if the value of some property is unchanged throughout a reaction, the quantity is said to be conserved.

cosmic rays: high-energy particles and atomic nuclei coming from outer space.

CP: combination of parity—mirror symmetry—and charge (particle-antiparticle) symmetry.

cyclotron: early form of particle accelerator.

down quark: lightest *quark* with electrical charge –1/3; constituent of *protons* and *neutrons*.

electromagnetic force: fundamental force that acts through forces between electrical charges and the magnetic force.

electron: lightweight electrically charged constituent of the *atom*.

electroweak force: theory uniting the *electromagnetic* and *weak forces*.

eV (electronvolt): unit of energy; the amount of energy that an *electron* gains when accelerated by one volt.

$E = mc^2$ (energy and mass units): technically the unit of MeV or GeV is a measure of the rest energy, $E = mc^2$, of a particle, but it is often traditional to refer to this simply as *mass*, and to express masses in MeV or GeV.

Fermilab: Fermi National Accelerator Laboratory, near Chicago, USA.

fermion: generic name for a particle with half-integer amount of *spin*, measured in units of *Planck's constant*. Examples are the *quarks* and *leptons*.

flavour: generic name for the qualities that distinguish the various *quarks* (*up*, *down*, *charm*, *strange*, *bottom*, *top*) and *leptons* (*electron*, *muon*, *tau*, *neutrinos*), thus flavour includes electric charge and *mass*.

gamma ray: *photon*; very high-energy electromagnetic radiation.

generation: *quarks* and *leptons* occur in three 'generations'. The first generation consists of the *up* and *down quarks*, the *electron* and a *neutrino*. The second generation contains the *charm* and *strange quark*, the *muon*, and another *neutrino*, while the third, and most massive, generation contains the *top* and *bottom quarks*, the *tau*, and a third variety of *neutrino*. We believe that there are no further examples of such generations.

GeV: unit of energy equivalent to a thousand million (10^9) *eV*.

gluon: massless particles that grip *quarks* together making *hadrons*; carrier of the *QCD* forces.

hadron: particle made of *quarks* and/or antiquarks, which feels the strong interaction.

Higgs boson: massive particle, discovered in 2012, which is the source of *mass* for fundamental particles such as the *electron*, *quarks*, *W* and *Z bosons*.

ion: *atom* carrying electric charge as a result of being stripped of one or more *electrons* (positive ion), or having an excess of *electrons* (negative ion).

K (kaon): variety of strange *meson*.

keV: a thousand *eV*.

kinetic energy: the energy of a body in motion.

LEP: Large Electron Positron collider at CERN. LEP closed in 2000 to make way for the LHC.

lepton: particles such as the *electron* and *neutrino* that do not feel the *strong force* and have *spin* 1/2.

LHC: Large Hadron Collider; accelerator at CERN.

LHCb: detector at the LHC, designed to be sensitive to particles containing bottom quarks.

linac: abbreviation for linear accelerator.

MACHO: acronym for Massive Compact Halo Object.

magnetic moment: quantity that describes the reaction of a particle to the presence of a magnetic field.

mass: the inertia of a particle or body, and a measure of resistance to acceleration; note that your 'weight' is the force that gravity exerts on your mass, so you have the same mass whether on Earth, on the Moon, or in space, even though you may be 'weightless' out there.

meson: class of *hadron*; made of a single *quark* and an antiquark.

MeV: a million *eV*.

meV: a millionth of an *eV*.

microsecond: one millionth of a second.

molecule: a cluster of *atoms*.

muon: heavier version of the *electron*.

nanosecond: one billionth of a second.

neutrino: electrically neutral particle, member of the *lepton* family; feels only the weak and gravitational forces.

neutron: electrically neutral partner of a *proton* in the atomic nucleus which helps stabilize the nucleus.

parity: the operation of studying a system or sequence of events reflected in a mirror.

photon: massless particle that carries the electromagnetic force.

picosecond: one millionth of a millionth of a second.

pion: the lightest example of a *meson*; made of an *up* and/or *down* flavour of *quark* and antiquark.

Planck's constant (h): a very small quantity that controls the workings of the universe at distances comparable to, or smaller than, the size of *atoms*. The fact that it is not zero is ultimately the reason why the size of an atom is not zero, why we cannot simultaneously know the position and speed of an atomic particle with perfect precision, and why the quantum world is so bizarre compared to our experiences in the world at large. The rate of *spin* of a particle is also proportional to h (technically, to units or half-integer units of h divided by 2π).

positron: antiparticle of an *electron*.

proton: electrically charged constituent of the atomic nucleus.

QCD (quantum chromodynamics): theory of the *strong force* that acts on *quarks*.

QED (quantum electrodynamics): theory of the *electromagnetic force*.

QFD (quantum flavour dynamics): unified theory of the weak and electromagnetic forces.

quarks: seeds of *protons*, *neutrons*, and *hadrons*.

radioactivity: see *beta decay*.

SLAC: Stanford Linear Accelerator Center, California, USA.

SNO: Sudbury Neutrino Observatory, underground laboratory in Sudbury, Ontario.

spark chamber: device for revealing the passage of electrically charged particles.

spin: measure of rotary motion, or intrinsic *angular momentum*, of a particle; measured in units of *Planck's constant*.

strange particles: particles containing one or more *strange quarks* or antiquarks.

strange quark: quark with electrical charge –1/3, more massive than the *down quarks* but lighter than the *bottom quark*.

strangeness: property possessed by all matter containing a *strange quark* or antiquark.

strong force: fundamental force, responsible for binding *quarks* and antiquarks to make *hadrons*, and gripping *protons* and *neutrons* in atomic nuclei; described by *QCD* theory.

Superkamiokande: underground detector of *neutrinos* and other particles from *cosmic rays*, located in Japan.

SUSY (supersymmetry): theory uniting *fermions* and *bosons*, where every known particle is partnered by a particle yet to be discovered whose *spin* differs from it by one half.

symmetry: if a theory or process does not change when certain operations are performed on it, then we say that it possesses a symmetry with respect to those operations. For example, a circle remains unchanged after rotation or reflection; it therefore has rotational and reflection symmetry.

synchrotron: modern circular accelerator.

tau: heavier version of the *muon* and *electron*.

top quark: the most massive *quark*; has charge +2/3.

UA1 and UA2: detectors at the CERN proton-antiproton collider, which discovered the W and Z bosons in 1983.

unified theories: attempts to unite the theories of the *strong*, *electromagnetic*, and *weak forces*, and ultimately gravity.

up quark: *quark* with an electrical charge of +2/3; constituent of *protons* and *neutrons*.

W boson: electrically charged massive particle, carrier of a form of the *weak force*; sibling of the *Z boson*.

weak force: fundamental force, responsible for *beta decay*; transmitted by *W* or *Z bosons*.

WIMP: acronym for 'weakly interacting massive particle'.

Z boson: electrically neutral massive particle, carrier of a form of the *weak force*; sibling of the *W boson*.

Index

For the benefit of digital users, indexed terms that span two pages (e.g., 52–53) may, on occasion, appear on only one of those pages.

E